GETTING STARTED WITH FPGAs

GETTING STARTED WITH FPGAs

Digital Circuit Design, Verilog, and VHDL for Beginners

by Russell Merrick

no starch
press®

San Francisco

Printed in the United States of America

Second printing

28 27 26 25 24   2 3 4 5 6

ISBN-13: 978-1-7185-0294-9 (print)
ISBN-13: 978-1-7185-0295-6 (ebook)

Published by No Starch Press®, Inc.
245 8th Street, San Francisco, CA 94103
phone: +1.415.863.9900
www.nostarch.com; info@nostarch.com

Publisher: William Pollock
Managing Editor: Jill Franklin
Production Manager: Sabrina Plomitallo-González
Production Editor: Jennifer Kepler
Developmental Editor: Nathan Heidelberger
Cover Illustrator: Gina Redman
Interior Design: Octopod Studios
Technical Reviewer: Mike D. Smith
Copyeditor: Rachel Head
Proofreader: James M. Fraleigh

Figure 9-3 was re-created based on an image by Mike Toews, used under the CC BY-SA 3.0 license, *https://commons.wikimedia.org/wiki/File:Analog_digital_series.svg.*

Library of Congress Cataloging-in-Publication Data

```
Names: Merrick, Russell, author.
Title: Getting started with FPGAs : digital circuit design, Verilog, and VHDL for beginners /
  by Russell Merrick.
Description: San Francisco, CA : No Starch Press, [2024] | Includes index.
Identifiers: LCCN 2023011303 (print) | LCCN 2023011304 (ebook) | ISBN 9781718502949 (paperback) |
  ISBN 9781718502956 (ebook)
Subjects: LCSH: Field programmable gate arrays. | Digital electronics. | Verilog (Computer
  hardware description language). | VHDL (Computer hardware description language).
Classification: LCC TK7895.G36 M47 2024 (print) | LCC TK7895.G36  (ebook) |
  DDC 621.39/5–dc23/eng/20230419
LC record available at https://lccn.loc.gov/2023011303
LC ebook record available at https://lccn.loc.gov/2023011304
```

[S]

For everyone who has supported me
on this journey.

About the Author

Russell Merrick grew up near Boston, Massachusetts, and graduated from the University of Massachusetts in 2007 with a degree in electrical engineering. Later he earned a master's in electrical engineering. He has worked in the defense industry at BAE Systems and L-3 Communications, in aerospace at a satellite propulsion startup called Accion Systems, and most recently in commercial electronics at a fitness wearable company called WHOOP. Russell has been creating content for FPGA developers at *https://nandland.com* and the accompanying YouTube channel since 2014, helping them get a solid footing in their digital design pursuits. He has three children and one very supportive wife.

About the Technical Reviewer

Mike D. Smith is the founder and principal designer at Bold Circuits LLC, which provides full-service electronic design services to the commercial, military, and research markets. He has over 20 years of industry experience and specializes in system-level electronics design, high-speed digital design, PCB layout, and FPGA development. Mike has a BS in electrical engineering and a BA in German from the University of Rhode Island and an MS in computer engineering from the University of Massachusetts, with a computing and embedded systems hardware and architecture concentration. He teaches a senior course in practical PCB design at the University of Rhode Island and serves as a consulting technical director to their ELECOMP Capstone Design Program, providing technical advisory services to industry-sponsored senior design teams.

BRIEF CONTENTS

CONTENTS IN DETAIL

ACKNOWLEDGMENTS

There are so many people who have helped make this book a reality in small and large ways. Throughout my career, I've had mentors who have guided me and pushed me to become better at my craft. There's nothing more rewarding than working with someone who is smarter than you are, so go find people who make you feel stupid in comparison. Try to glean anything you can from them. Look at their code, ask dumb questions, and you'll improve.

I'd like to thank everyone who has supported me in my Nandland pursuits—those who have liked, commented, or subscribed to videos on my YouTube channel; purchased a Go Board; or supported my Patreon. You've built up my confidence and helped me stay motivated to keep cranking out FPGA content for years and years.

I never imagined that writing a book would be as lengthy a process as it turned out to be. Many thanks to my editor, Nathan Heidelberger, who has had the biggest impact on the quality of this book. Nathan doesn't have an engineering background, but he was able to give me valuable feedback from an outsider's perspective when concepts were unclear. Similarly, I'm grateful to my technical reviewer, Mike D. Smith, a coworker and friend from my BAE Systems days. His attention to the technical explanations ensured that this book would be accessible and, equally importantly, accurate.

Finally, I'd like to thank my family for allowing me the (significant) time required to write this book. Many nights and weekends have been impacted, so to my wife, Christine, I would like to say thank you for always being supportive and encouraging. I couldn't have done this without your help.

INTRODUCTION

In my first job out of college as an entry-level electrical engineer, I once worked on an old design that had a timer circuit on it. Using a simple resistor and capacitor, the circuit would wait for 50 milliseconds (ms) to elapse, then trigger an action. We needed to change that 50 ms timer to 60 ms, but this small change required a monumental effort: we would have to physically remove the capacitors and resistors from hundreds of circuit boards and replace them with new ones.

Fortunately, we had a team of field programmable gate array (FPGA) designers who came to the rescue. With their help, we were able to implement the same functionality inside an FPGA. Then, in a matter of minutes, we could change code to set the timer to any arbitrary value we wanted, without having to touch a soldering iron. This faster pace of progress excited me, and I quickly got hooked on FPGAs.

Eventually I transitioned to working with FPGAs full-time, and it was around then that I started reading and responding to FPGA-related questions on Stack Overflow. Often these questions came from FPGA beginners who were confused about basic concepts. I saw the same types of questions asked again and again, and realized there wasn't a single place where people could learn about FPGAs in a simple, easy-to-understand way. Sure, there were many online references for Verilog and VHDL, the two most popular FPGA programming languages, but there was relatively little information on what those languages were actually *doing*. What components are really being created within your FPGA when you write a certain line of code? How are things being wired up? What does it mean to run operations in parallel versus serially?

Rather than continuing to answer the same questions over and over, I started my own website, *https://nandland.com*, where I began writing longer articles about FPGAs. As traffic grew, I started making YouTube videos as well. I even created my own FPGA development board to provide hands-on experience for beginners. In all these endeavors, my goal has been to demystify FPGAs while making the information approachable and engaging for those just starting out. Writing this book has allowed me to delve even deeper into the subject, to build a solid foundation for anyone interested in exploring the exciting world of FPGA design.

Who Is This Book For?

I've tried to make this book as accessible as possible so that a broad range of people will be able to read and understand the material. The intended audience is anyone who is curious about how digital programmable logic works and how FPGAs can be used to solve a wide variety of problems. Maybe you're a college student who encountered FPGAs in a class and was left intrigued but confused, or someone in the electronics industry who has been exposed to FPGAs at work. Perhaps you're a tinkerer or hardware hacker, or a software developer interested in programming at a much lower level than you're used to. This book is very approachable for all of these groups.

I'm assuming you've had at least some exposure to a conventional programming language, like Python, C, or JavaScript. It will be helpful if you understand concepts like functions, conditional statements (if...else), loops, and other basic programming techniques. You don't need any prior experience with Verilog or VHDL, however; this book will introduce the basics of these languages.

FPGAs lie at the intersection of hardware and software, so having some interest in electronics is helpful. We'll sometimes discuss concepts like voltage and current within the FPGA. Here again, it will be useful if you've had some basic introduction to these terms, but it won't be required to get value out of those sections.

What This Book Isn't

This book isn't intended to teach you every facet of Verilog or VHDL. As I said earlier, there are many online resources if that's your goal. Instead, my aim is to teach you how FPGAs work so you can understand what your Verilog or VHDL is doing, and thus make more intelligent choices about your designs. That said, we *will* look at a large amount of code throughout the book. All of it is thoroughly explained, so you won't need prior experience with these programming languages to follow along. You'll gain a strong base of Verilog and VHDL knowledge as you read, and the confidence to augment that knowledge through independent study.

The book includes various projects that you'll be able to carry out on real hardware using the iCE40 line of FPGAs from Lattice Semiconductor. I've focused on these comparatively cheap, simple FPGAs to make the hands-on parts of the book as accessible as possible. More expensive FPGAs have many extra bells and whistles; they're very cool, but they can be overwhelming for beginners. iCE40 FPGAs are still highly capable, but with fewer of these high-end features available. As such, this book won't explore sophisticated features like SerDes and hard-core processors in a hands-on way, nor will we dwell on the more complicated FPGA tools required to use them. We *will* discuss some of these features at a high level, however, so you'll gain the background knowledge to work with them if you choose to upgrade to a fancier FPGA.

What's in the Book?

This book combines high-level discussion, detailed code examples, and hands-on projects. Each code listing is shown in both Verilog and VHDL, so whichever language you want to use for FPGA development, you'll be able to follow along. There's also an extensive glossary at the end of the book for your reference. Here's what you'll find in each chapter:

Chapter 1: Meet the FPGA   Introduces FPGAs and talks about their strengths and weaknesses. Being an engineer is about knowing which tool to use in which scenario. Understanding when to use an FPGA—and when not to—is crucial.

Chapter 2: Setting Up Your Hardware and Tools   Gets you set up with the Lattice iCE40 series of FPGAs. You'll download and install the FPGA tools and learn how to run them to program your FPGA.

Chapter 3: Boolean Algebra and the Look-Up Table   Explores one of the two most fundamental FPGA components: the look-up table (LUT). You'll learn how LUTs perform Boolean algebra and take the place of dedicated logic gates.

Chapter 4: Storing State with the Flip-Flop   Introduces the second fundamental FPGA component: the flip-flop. You'll see how flip-flops store state within an FPGA, giving the device memory of what happened previously.

Chapter 5: Testing Your Code with Simulation   Discusses how to write testbenches to simulate your FPGA designs and make sure they work correctly. It's hard to see what's going on inside a real physical FPGA, but simulations let you investigate how your code is behaving, find bugs, and understand strange behaviors.

Chapter 6: Common FPGA Modules   Shows how to create some basic building blocks common to most FPGA designs, including multiplexers, demultiplexers, shift registers, and first in, first out (FIFO) and other memory structures. You'll learn how they work and how to combine them to solve complex problems.

Chapter 7: Synthesis, Place and Route, and Crossing Clock Domains Expands on the FPGA build process, with details about synthesis and the place and route stage. You'll learn about timing errors and how to avoid them, and how to safely cross between clock domains within your FPGA design.

Chapter 8: The State Machine   Introduces the state machine, a common model for keeping track of the logical flow through a sequence of events in an FPGA. You'll use a state machine to implement an interactive memory game.

Chapter 9: Useful FPGA Primitives   Discusses other important FPGA components besides the LUT and the flip-flop, including the block RAM, the DSP block, and the phase-locked loop (PLL). You'll learn different strategies for harnessing these components and see how they solve common problems.

Chapter 10: Numbers and Math   Outlines simple rules for working with numbers and implementing math operations in an FPGA. You'll learn the difference between signed and unsigned numbers, fixed-point and floating-point operations, and more.

Chapter 11: Getting Data In and Out with I/O and SerDes   Examines the input/output (I/O) capabilities of an FPGA. You'll learn the pros and cons of different types of interfaces and be introduced to SerDes, a powerful FPGA feature for high-speed data transmission.

Appendix A: FPGA Development Boards   Suggests some FPGA development boards that you can use for this book's projects.

Appendix B: Tips for a Career in FPGA Engineering   Outlines strategies for finding an FPGA-related job, in case you want to pursue FPGA design professionally. I'll make suggestions on how to build a good resume, prepare for interviews, and negotiate for the best-possible job offer.

What You'll Need

Although not strictly required, I recommend having a development board with a Lattice iCE40 FPGA so you can complete the book's hands-on projects. There's nothing more satisfying than learning about a concept and then being able to implement that concept on real hardware. Chapter 2 discusses what to look for in a development board and exactly what you'll need for the book's projects in more detail. Briefly, the development board should have a USB connection and peripherals like LEDs, push-button switches, and a seven-segment display. Appendix A describes some development boards that will work.

The software tools for working with iCE40 FPGAs run best on Windows. If you don't have a Windows computer, I recommend running the tools inside a Windows virtual machine. We'll discuss installing these tools in Chapter 2.

Online Resources

The code presented in this book is available online via a GitHub repository. You can access it at *https://github.com/nandland/getting-started-with-fpgas*. You'll also find more information and FPGA project ideas online at *https://nandland.com*.

1

MEET THE FPGA

An *FPGA*, short for *field programmable gate array*, is a highly capable type of *integrated circuit*, an electronic circuit in a single package. The *field programmable* part of the name indicates that FPGAs can be reprogrammed when in the field (that is, without having to return them to the manufacturer). The *gate array* part indicates that an FPGA is made up of a two-dimensional grid featuring a large number of gates, fundamental units of digital logic that we'll discuss in depth in Chapter 3.

The name is actually a bit of an anachronism. The reality is that some FPGAs aren't field programmable, and most are no longer just an array of simple gates. In fact, they're much more sophisticated than that. Despite these exceptions, the name has stuck over the years, and it highlights a unique characteristic of FPGAs: their incredible flexibility. An FPGA's uses are limited only by the designer's imagination. Other digital programmable devices, such as microcontrollers, are designed with a specific set of capabilities; you can only do something if that feature is built in. By contrast, an FPGA's array of gates (or the more modern equivalent) is like a blank slate that you can program, and reprogram, and reprogram to do almost anything you want, with fewer restrictions. This freedom doesn't come without trade-offs, however, and FPGA development demands a unique set of skills.

Learning how to work with FPGAs requires a different style of thinking from traditional computer programming. Traditional software engineering, like programming in C, for example, is serial: first this happens, then this happens, and finally this happens. This is because C is compiled to run on a single processor, or CPU, and that CPU is a serial machine. It processes one instruction at a time.

FPGAs, on the other hand, work in parallel: everything is happening at the same time. Understanding the difference between serial and parallel programming is fundamental to working with FPGAs. When you can think about solving a problem using parallel methods, your overall problem-solving skills will increase. These skills will also translate to other, non-FPGA applications; you'll begin to see problems differently than if you were only thinking about them serially. Learning how to think in parallel rather than serially is a critical skill for becoming an FPGA engineer, and it's one you'll develop throughout this book.

FPGAs are a lot of fun to work with. When you create an FPGA design using Verilog or VHDL (more on these languages later in this chapter), you're writing code at the lowest possible level. You're literally creating the physical connections, the actual wires, between electrical components and input/output pins on your device. This allows you to solve almost any digital problem: you have complete control. It's a much lower level of programming than working with a microcontroller that has a processor, for example. For this reason, learning about FPGAs is an excellent way to become familiar with hardware programming techniques and better understand how exactly digital logic works in other applications. You'll gain a newfound respect for the complexities of even the simplest integrated circuits once you start working with FPGAs.

This first chapter sets you up for diving deeper into FPGAs by providing some background information. We'll briefly cover the history of FPGAs, from their initial creation in the 1980s through today, and explore some of their common uses. We'll also consider how FPGAs compare to other common digital components, such as microcontrollers and application-specific integrated circuits (ASICs). Finally, we'll discuss the differences between Verilog and VHDL, the two most popular languages for working with FPGAs.

A Brief History of FPGAs

The very first FPGA was the XC2064, created by Xilinx in 1985. It was very primitive, with a measly 800 gates, a fraction compared to the millions of gate operations that can be performed on today's FPGAs. It was also relatively expensive, costing $55, which adjusted for inflation would be around $145 today. Still, the XC2064 kicked off an entire industry, and (alongside Altera) Xilinx has remained one of the dominant companies in the FPGA market for more than 30 years.

Early FPGAs like the XC2064 were only able to perform very simple tasks: Boolean operations such as taking the logical OR of two input pins and putting the result onto an output pin (you'll learn much more about Boolean operations and logic gates in Chapter 3). In the 1980s, this type of problem required a dedicated circuit built of OR gates. If you also needed to perform a Boolean AND on two different pins, you might have to add another circuit, filling up your circuit board with these dedicated components. When FPGAs came along, a single device could replace many discrete gate components, lowering costs, saving component space on the circuit board, and allowing the design to be reprogrammed as the requirements of the project changed.

From these humble beginnings, the capabilities of FPGAs have increased dramatically. Over the years, the devices have been designed with more *hard intellectual property (IP)*, or specialized components within the FPGA that are dedicated to performing a specific task (as opposed to *soft* components that can be used to perform many tasks). For example, hard IP blocks in modern FPGAs let them interface directly with USB devices, DDR memory, and other off-chip components. Some of these capabilities (like a USB-C interface) would simply not be possible without some dedicated hard IP to do the job. Companies have even placed dedicated processors (called *hard processors*) inside FPGAs so that you can run normal C code within the FPGA itself.

As the devices have evolved, the FPGA market has undergone many mergers and acquisitions. In 2020, the chip-making company AMD purchased Xilinx for $35 billion. It's plausible that this purchase was a response to its main competitor Intel's 2015 acquisition of Altera for $16.7 billion. It's interesting that two companies focused predominantly on CPUs decided to purchase FPGA companies, and there's much speculation as to why. In general, it's thought that as CPUs mature, dedicating some part of the chip to FPGA-like reprogrammable hardware seems to be an idea worth pursuing.

Apart from Xilinx and Altera (which from here on I'll be calling by their parent company names, AMD and Intel, respectively), other companies have carved out their own niches within the FPGA market. For example, Lattice Semiconductor has done well for itself making mostly smaller, less expensive FPGAs. Lattice has been happy to play on its own in this lower end of the market, while letting AMD and Intel slug it out at the higher end. Today, the open source community has embraced Lattice FPGAs, which have been reverse-engineered to allow for low-level hacking. Another medium-sized player in the FPGA space, Actel, was acquired by Microsemi in 2010 for $430 million. Microsemi itself was acquired by Microchip Technology in 2018.

Popular FPGA Applications

In their modern, highly capable and flexible form, FPGAs are used in many interesting areas. For example, they're a critical component in the telecommunications industry, where they're often found in cell phone towers. They route internet traffic to bring the internet to your smartphone, allowing you to stream YouTube videos on your bus ride to work.

FPGAs are also widely used in the finance industry for high-frequency trading, where companies use algorithms to automatically buy and sell stocks incredibly quickly. Traders have found that if you can execute a stock purchase or sale slightly quicker than the competition, you can gain a financial edge. The speed of execution is paramount; a tiny bit of latency can cost a company millions of dollars. FPGAs are well suited to this task because they're very fast and can be reprogrammed as new trading algorithms are discovered. This is an industry where milliseconds matter, and FPGAs can provide an advantage.

FPGAs are used in the defense industry as well, for applications like radar digital signal processing. FPGAs can process received radar reflections using mathematical filters to see small objects hundreds of miles away. They're also used to process and manipulate images from infrared (IR) cameras, which can see heat rather than visible light, allowing military operatives to see people even in complete darkness. These operations are often highly math-intensive, requiring many multiplication and addition operations to happen at parallel—something that FPGAs excel at.

Another area where FPGAs have found a niche is in the space industry: they can be programmed with redundancies to hedge against the effects of radiation bombardment, which can cause digital circuits to fail. On Earth, the atmosphere protects electronics (and people) from lots of solar radiation, but outer space doesn't have that lovely blanket, so the electronics on satellites are subjected to a much harsher environment.

Finally, FPGAs are also getting interest from the artificial intelligence (AI) community. They can be used to accelerate neural nets, another massively parallel computational problem, and thus are helping humans attack issues that weren't solvable using traditional programming techniques: image classification, speech recognition and translation, robotics control, game strategy, and more.

This look at popular FPGA applications is far from exhaustive. Overall, FPGAs are a good candidate for any digital electronics problem where high bandwidth, low latency, or high processing capability is needed.

Comparing Common Digital Logic Components

Despite how far they've come since their early days and the wide range of applications they're used for, FPGAs are still a relatively niche technology compared to other digital logic components like microcontrollers and ASICs. In this section, we'll compare these three technologies. You'll see why FPGAs are a good solution for some problems but not others, and how they have tough competition from other devices, especially microcontrollers.

FPGAs vs. Microcontrollers

Microcontrollers are everywhere. If you aren't an embedded software engineer, you may not realize how many toys, tools, gadgets, and devices are controlled with small and inexpensive microcontrollers: everything from TV remote controls to coffee makers to talking toys. If you're an electronics hobbyist, you might be familiar with the *Arduino*, which is powered by a small microcontroller from Atmel (now Microchip Technology, the same company that owns what used to be Actel). Millions of Arduinos have been sold to hobbyists around the world. They're cheap, fun, and relatively easy to work with.

So why are microcontrollers everywhere, but not FPGAs? Why isn't there an FPGA controlling your coffee maker or making your Elmo doll come to life? The main reason is cost. The consumer electronics industry, which uses the largest number of microcontrollers overall, is incredibly sensitive to cost. Consumers like you and I want the least expensive products we can possibly buy, and the companies that make those products will shave off every penny possible to make that happen.

Microcontrollers come in seemingly endless varieties, with each one designed for a very specific purpose. This helps companies drive costs down. For example, if your product needs one analog-to-digital converter (ADC), two USB interfaces, and at least 30 general purpose input/output (GPIO) pins, there's a microcontroller with exactly those specifications. What if you realize you only need one USB interface? There's probably a different microcontroller with those specifications, too. With such variety, there's no need to pay for extra features. Companies can find a microcontroller with the bare minimum of what they need, and save money in the process.

FPGAs, on the other hand, are much more general. With a single FPGA, you might create five ADC interfaces and no USB interface, or three USB and no ADC interfaces. You pretty much have a blank slate at your disposal. As you'll learn, however, FPGAs need to have many internal wires (called *routing*) to support all these different possibilities, and all that routing adds cost and complexity. In many cases, you'll end up paying more for extra features and flexibility that you don't need.

Another contributing factor to cost is quantity. If you buy 10 million microcontrollers, which is not unrealistic in the field of consumer electronics, you'll pay less per chip than you would if you only bought 100,000. FPGAs, meanwhile, are typically produced and sold in relatively low quantities, so they cost more per unit. It's a bit of a chicken-and-egg situation, where FPGAs could be less expensive if there were more of them, but for there to be more of them, they would have to be less expensive. If the costs were the same as with microcontrollers, would there be more FPGAs in use? I think it's likely that there would be, but FPGAs are also more complicated to use, so that works against them as well.

Since microcontrollers are designed for specific purposes, they can be very easy to set up. You can get a basic design up and running on a microcontroller in a few hours. By contrast, you need to program *everything* inside the FPGA, and this is very time-consuming. Although there are some hard IP blocks to get you started, the majority of the device is programmable logic—that blank slate we talked about—that you need to design yourself. Writing all the code to do what you need also takes longer in a language like Verilog or VHDL than C, which is commonly used to program microcontrollers. With C, you're writing code at a higher level, so you can do more with a single line. With Verilog and VHDL you're writing at a much lower level: individual gates and wires are literally being created with your code. You can think of low-level programming like working with individual LEGO bricks and high-level programming like working with preconstructed LEGO sets. This adds complexity, which adds time, which also increases costs. Engineers want the simplest solution, and most often a microcontroller is simpler than an FPGA.

Another factor to consider is how much power the device consumes. Many electronic devices run off batteries, and it's critical to maximize their lifetime by making the devices as low-power as possible. The more power they use, the more often you'll have to change the batteries, which is something nobody wants to do. Again, since a microcontroller is designed for a specific use, it can be optimized to draw incredibly little power, enabling a single AAA battery to power a Bluetooth mouse for months, for example. FPGAs, with all their routing resources, are simply unable to compete with microcontrollers in terms of power consumption. That's not to say you can't use an FPGA in a battery-powered application, but head-to-head the microcontroller will win that battle every time.

Summarizing, microcontrollers almost always dominate in terms of cost, ease of use, and power consumption. So why would anyone use an FPGA over a microcontroller? There are other factors to consider, such as speed and flexibility, and here the tables turn in favor of the FPGA.

When I say speed, I mean two things: bandwidth and computations. *Bandwidth* is the rate of data transfer across a path. FPGAs can have incredibly large bandwidth, much more than any microcontroller could ever attain. They can process hundreds of gigabits per second with no trouble at all. This might be useful, for example, when driving multiple 4K displays. FPGAs are often used in video editing hardware that requires enormous amounts of bandwidth to keep up with the data streams. Their high bandwidth allows them to move tremendous amounts of data from various external interfaces (USB-C, Ethernet, ADCs, memories, and more) at very fast rates.

As for computational speed, the number of mathematical computations an FPGA can perform in a second dwarfs anything a microcontroller can do. A microcontroller usually has just one processor, and with all computations going through the same processor, the number of computations that can be performed each second is limited. An FPGA, on the other hand, can run many computations in parallel. For example, you can run hundreds of multiplication operations at the same time, something that simply isn't possible with a microcontroller. This might be useful when running large mathematical filters on data, which often involve many multiplication and addition operations occurring at very fast rates.

The other major benefit of an FPGA is its flexibility. I said microcontrollers come in endless varieties, but of course, this is a slight exaggeration. If your design has some particularly exotic requirement—say, if it needs 16 ADC interfaces—there might not be any microcontroller in the world that will meet your needs. FPGAs are much less limited. As I've mentioned, an FPGA is like a blank slate that can be programmed to do almost anything, providing you with tremendous flexibility to tackle a wide range of digital logic problems.

When you have an engineering problem, you need to choose the best tool possible to solve it. Often a microcontroller works very well, but occasionally it simply won't work due to speed or flexibility issues. In those situations, an FPGA is a good candidate. However, there's also another kind of device worth considering: an ASIC.

FPGAs vs. ASICs

An ASIC is a type of integrated circuit designed for a particular use. Unlike an FPGA, which can suit any number of uses, an ASIC is designed to be really good at one thing. You might think that it would always be better to have the flexibility of an FPGA, but there are trade-offs to consider. We've already compared FPGAs to microcontrollers in terms of cost, ease of use, power, speed, and flexibility. Let's now compare FPGAs and ASICs along those same lines.

ASICs are incredibly expensive to make in low quantities because they have a large *nonrecurring engineering (NRE)* cost: you need to pay a lot of money up front to a semiconductor foundry (or *fab*) to get that first ASIC chip. Often the NRE on an ASIC design can run into the millions of dollars. Whether or not you choose to design an ASIC highly depends on how many chips you'll need. If you're making low quantities of something, even into the tens of thousands, it's unlikely you'll ever be able to recover the upfront cost of an ASIC. If you need millions of chips, however, then an ASIC starts to become an attractive option, since each chip after the first one is very cheap (often under $1). Compare that to an FPGA, where a single chip often costs more than $10, and you start to see that FPGAs just don't make financial sense in large volumes. Generalizing, in smaller quantities FPGAs usually win against ASICs, but in larger quantities they can't compete as well.

An area where FPGAs always win out over ASICs is in terms of ease of use. The process of designing an ASIC is very complicated. Plus, you need to make sure your design is free of bugs *before* you go to the fab to make the chip, or you'll have wasted your NRE. Most FPGAs, on the other hand, can be fixed in the field (hence *field programmable*), so even if you find a bug after shipping your product to customers, you can update the code and remedy the issue. That's simply not possible with an ASIC. Therefore, you must spend significant engineering time and effort verifying that your ASIC design is as bug-free as possible before getting it fabricated. There's actually an entire discipline called *verification engineering* that does just this, something we'll explore in more detail in Chapter 5.

One large benefit to an ASIC is that it can be optimized for low power. ASICs are finely tuned for their specific application; they have what they need and no more. Meanwhile, recall that an FPGA has a significant number of wires and interconnections, which give it its flexibility but mean it uses more power. A further advantage ASICs have over FPGAs is that they can use fabrication techniques that optimize them for low power at the transistor level. For a real-life example, when Bitcoin was new, people were using their home computers (CPUs) to mine it. This draws a lot of power per Bitcoin mined. Eventually, people realized that FPGAs could be programmed to mine Bitcoins, using less power than CPUs. Electrical power is expensive, so mining Bitcoins with FPGAs was more profitable. Further down the line, people realized that ASICs could mine Bitcoins even faster and using even less power than FPGAs. It became worth the up-front cost to create an ASIC dedicated to Bitcoin mining, because the cost savings from the lower power consumption were so significant. Today, Bitcoins are mined almost exclusively with ASICs.

When it comes to speed, FPGAs and ASICs both have large bandwidth and can move lots of data around. They're also both very capable at math operations, particularly multiplication and addition, and both can do those operations in parallel. ASICs have a small edge in this category: because they're built specifically for one purpose, they can often run a bit faster than FPGAs.

FPGAs, on the other hand, offer significantly more flexibility than ASICs. Having flexibility in what your design can do is very valuable, especially if you're working on a project that isn't clearly defined. Unlike ASICs, which are fixed, FPGAs can be reprogrammed over and over again, with features and functionality added or removed. Additionally, ASICs take a long time to design, make, and verify, but you can get started with an FPGA right away, so the speed of progress can be faster.

In general, ASICs win against FPGAs on cost when the volumes are very high, but not when they're low. They beat FPGAs on power consumption and have a slight edge on speed, but they lose to FPGAs on flexibility and ease of use. In reality, though, ASICs and FPGAs often go hand in hand. When a company wants to design an ASIC, it will typically start by designing a prototype with an FPGA, and then produce the ASIC. This method allows engineers to work with the hardware sooner and gain confidence in the product, prior to spending millions of dollars on a custom chip. The engineers can work through the bugs in the Verilog or VHDL code using the FPGA prototype, and fix issues when it's inexpensive and simpler to do so. This code isn't throwaway code either, since the same Verilog or VHDL used on an FPGA can be used to create an ASIC.

FPGAs vs. Microcontrollers vs. ASICs

That was a lot to take in, so let's briefly summarize what we've just discussed about FPGAs, microcontrollers, and ASICs. Table 1-1 provides an overview of where each type of device sits on the scale across different parameters. There are always exceptions, but the table provides a good generalization.

Table 1-1: Comparing an FPGA vs. a Microcontroller vs. an ASIC

	FPGA	Microcontroller	ASIC
Cost (low quantities)	Moderate	Cheap	Expensive
Cost (high quantities)	Moderate	Cheap	Cheap
Speed	Fast	Moderate	Fast+
Power	Moderate	Low	Low
Flexibility	High	Low	None
Ease of use	Medium	Easy	Difficult

Cost is often the dominant factor in why a microcontroller or an ASIC is chosen over an FPGA for large-volume applications. They're simply cheaper, and that counts in industries that are highly sensitive to cost. When performance is the most important consideration and the high initial cost and level of complexity are acceptable, ASICs are often preferred. The sweet spot for FPGAs is applications that are low volume but require high speed in terms of either bandwidth or computations, or require a very flexible and unique design (like having 16 ADCs to interface to).

Ultimately, FPGAs, microcontrollers, and ASICs are three tools in the engineer's toolbox. When looking at the requirements of your particular problem, you'll need to decide which of these tools provides the best solution. You wouldn't use a hammer to turn a screw; knowing which tool to use for which application is critical to becoming a strong engineer. The act of examining multiple possibilities and selecting a technical solution is often referred to in engineering as a *trade study*.

I love FPGAs, but when I look at a technical problem, often the right solution is a microcontroller: they're easy to use and inexpensive. A microcontroller isn't *always* the right solution, though. Sometimes you need more speed, or the problem is one that microcontrollers just aren't designed for. Selecting the right tool for your problem will make solving the problem much more enjoyable. And as you'll hopefully see, working with FPGAs is certainly enjoyable! It's like working with LEGO or building a house in Minecraft. Using simple, low-level building blocks, you can create something wonderfully complex.

Verilog and VHDL

As I've mentioned, there are two main languages for working with FPGAs (and ASICs): Verilog and VHDL. These languages are known as *hardware description languages (HDLs)* because they're used to define the behavior of digital logic circuits. Though syntactically Verilog and VHDL may look similar to traditional programming languages, it's important to realize that

HDLs are a different beast entirely. When you write your FPGA code with an HDL, you're working directly with wires, logic gates, and other discrete resources on the FPGA, whereas when you code with a traditional programming language you don't have that same low level of control over your device. Understanding the logic behind your Verilog or VHDL code and knowing what FPGA components you're instantiating with that code is a critical skill for a digital designer, and something we'll return to throughout the book.

FPGA beginners often wonder if it's better to learn Verilog or VHDL. There's no right answer; it really depends on your situation. To help you make that decision, let's compare and contrast the two languages to identify some reasons why you might select one over the other.

NOTE *All the code examples in this book will be shown in both Verilog and VHDL, back-to-back, so whichever language you choose, you'll be able to follow along.*

VHDL stands for VHSIC Hardware Description Language, and VHSIC, the acronym within an acronym, stands for Very High-Speed Integrated Circuit. In full, VHDL is the *Very High-Speed Integrated Circuit Hardware Description Language*—quite a mouthful! It was developed by the United States Department of Defense (DoD) in 1983 and borrows many features and syntax from another DoD-developed language called Ada. VHDL, like Ada, is strongly typed.

If you've never worked with a strongly typed language, it can be a bit challenging at first. Strong typing forces the designer to be very explicit with their code. For example, in a weakly typed language like Python or C, you can add a variable defined as an integer to a variable defined as a float without any problem. A strongly typed language like VHDL, however, would never allow something like this. When adding two numbers in VHDL, their widths (number of bits) and types need to match exactly, or the syntax checker will throw some cryptic error. Until you understand the strong type checking that the language is performing, it can be cumbersome to get what you need done as a beginner. When sorting out these issues in VHDL, you often need to create intermediary signals of the correct type, or use lots of type conversions throughout your code. This is one of several reasons VHDL often requires much more typing (on your keyboard, that is) than Verilog to perform the same functionality. If you want to use VHDL, it helps to be a fast typist.

Compared to VHDL, Verilog looks more similar to a software language like C, which makes it easier for some people to read Verilog code and understand what it's doing. Also, Verilog is weakly typed. It allows you to write code that's wrong, but more concise. It would have no problem adding a float to an integer, even if the result of that addition might be incorrect.

Another interesting point of comparison is that Verilog is case sensitive but VHDL is not. This means that a variable called RxDone isn't the same as a variable called rxDone in Verilog, but VHDL treats them as the same. It might seem odd that strongly typed VHDL isn't case sensitive while weakly typed Verilog is, but that's just the way history turned out. In my experience, Verilog's case sensitivity can create issues that are hard to diagnose. You might think two signals are the same, but the code sees them as different due to a capitalization discrepancy.

Ultimately, none of these points is the most important factor, though. You should choose between Verilog and VHDL based on which language you're more likely to use in school or at work. If your university uses Verilog, learn Verilog! If companies where you might want to work use VHDL, learn VHDL! The breakdown of who uses VHDL versus Verilog is highly dependent on where in the world you're living. If you compare VHDL and Verilog using Google Trends, you can start to get a pretty good idea of which language you should be learning first.

When you look at the overall search volumes for the two terms throughout the world, you'll find that "Verilog" tends to have more searches than "VHDL." Maybe that's because it's used more often, or maybe people have more trouble with Verilog and need to look up solutions online more than they do with VHDL. In any case, it's more revealing to break down the trends on a country-by-country basis. Figure 1-1 shows some examples.

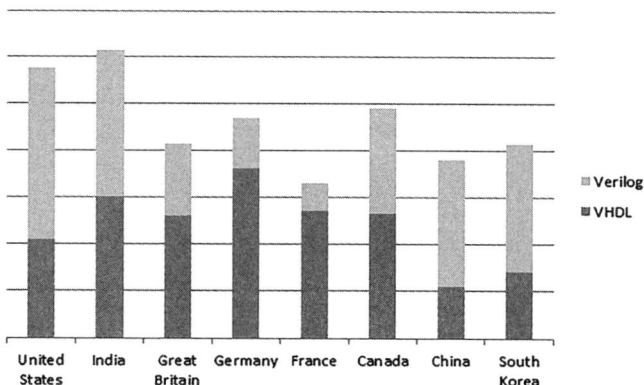

Figure 1-1: Verilog vs. VHDL search volumes in selected countries

India and the United States have the largest volumes of Google searches for the two terms, but while VHDL and Verilog appear to be roughly equal in popularity in India, Verilog is slightly more popular in the United States than VHDL. In fact, I know from personal experience that in the United States the defense industry favors VHDL, while the commercial industry favors Verilog. Notice that in Germany and France, VHDL is significantly more popular than Verilog. If you are from either of these two countries, I would highly recommend learning VHDL first! Conversely, in China and South Korea Verilog is much more popular than VHDL, so adjust your priorities accordingly.

In general, VHDL and Verilog are equally capable languages. You should choose which language to learn based on what suits your location and circumstances.

Summary

This chapter introduced you to FPGAs and provided you with an overview of their history and common applications. We compared FPGAs with micro-controllers and ASICs and saw where each type of integrated circuit really shines. You learned that FPGAs excel at applications that aren't cost sensitive, but where high speed, maximum flexibility, or unique interfaces are required. Finally, we looked at the two most popular hardware description languages used to work with FPGAs, Verilog and VHDL, and discussed how to choose the right language for you and your situation.

2

SETTING UP YOUR
HARDWARE AND TOOLS

This chapter walks you through the process of choosing an FPGA development board and setting up the associated software tools that you'll need to translate your Verilog or VHDL code into physical circuits on your FPGA. You'll learn about the features to look for in a board, download and install the tools you need to work with it, and test them out by designing your first FPGA project to target your development board. This project will also give you an overview of the main steps in the FPGA development process.

An FPGA development board isn't strictly required to use this book. You can still learn from the projects without a board, and you can always test your Verilog or VHDL code by running it through a free online FPGA simulator tool like EDA Playground (a topic we'll cover in Chapter 5).

However, there's something satisfying about writing some code, program-ming it to a development board, and seeing the results in action—even when it does something as simple as blinking an LED. For this reason, I highly recommend that you have an FPGA development board on hand when learning about FPGAs.

Choosing an FPGA Development Board

An FPGA development board (or *dev board*) is a printed circuit board (PCB) with an FPGA on it that allows you to program the FPGA with your Verilog or VHDL code and test it out. The board may also have peripherals on it that are connected to the FPGA, such as LEDs, switches, and connectors for linking the FPGA to other devices. FPGA development boards range from devices the size of a pack of gum that cost less than $100 to laptop-size devices that cost thousands of dollars. With such a wide range of options, there are many factors you should consider when choosing a development board, including price, ease of use, and enjoyability:

Cost

For an FPGA beginner, I recommend starting off with an inexpensive development board. Larger, more expensive boards often have many extra features, like SerDes and DDR memory, that are unnecessary and potentially overwhelming for new users. You can always invest in one of these more sophisticated boards as your skills mature and you grow out of your first board.

Simplicity

The board you start out with and the software required to work with it should be simple to use. It's challenging enough to learn how an FPGA works; if you also have to learn how to use a complicated design tool, the process becomes even more difficult. I recommend focusing on boards built around Lattice Semiconductor's iCE40 family of FPGAs, since these FPGAs are compatible with a lightweight and simple set of software tools: iCEcube2 and Diamond Programmer. These programs are streamlined to do the minimum required to build your FPGA, with-out all the bells and whistles of more advanced programs. You'll learn how to use both of them in this chapter.

Fun

An FPGA development board should be fun to use, with peripherals like LEDs, push buttons, and seven-segment displays that you can take advantage of in different projects. Some cheaper boards cut costs by removing peripherals; they just have an FPGA and nothing else. FPGA development is much more fun and interesting when you can interface the FPGA with other devices.

Keep these factors in mind as you consider the development boards available on the market.

Requirements for the Book

You'll get the most value out of this book if you follow along with the projects and program your own development board. To go through the projects exactly as written, you'll need your development board to have the following features (Appendix A lists a few boards that meet these requirements, or can meet them with a few modifications):

Lattice iCE40 FPGA

iCE40 FPGAs have emerged as the best option for FPGA beginners. They've been available for many years at affordable prices, while providing enough resources to support interesting projects. The iCE40 architecture is relatively simple, with few distracting bells and whistles, so you can focus on what's important. As I mentioned earlier, iCE40 FPGAs are compatible with the free, easy-to-use iCEcube2 and Diamond Programmer software tools, which we'll explore in this chapter. The iCE40 family is also compatible with open source FPGA tools, if you want to avoid proprietary software altogether.

USB

Your development board should have a USB interface to power and program the board. This way, all you need is one USB cable and you're ready to go. Older FPGA development boards often require an external programmer (a separate piece of hardware that can itself cost hundreds of dollars), so make sure simple built-in USB programming is possible with the board you choose.

LEDs

The book's projects assume that your board has four LEDs. These are a convenient way to get output from the FPGA. For example, our first project later this chapter will involve lighting up the LEDs, which allows you to get immediate feedback that you've successfully programmed the FPGA. There's nothing more satisfying than getting that first LED to light up!

Switches

For each of the four LEDs, you'll need a corresponding push-button switch. These switches provide input to the FPGA, allowing you to easily change the state of the board.

Seven-Segment Display

Your board will need one seven-segment display to implement the memory game project in Chapter 8. This kind of display provides a fun way to output data. Lighting up individual LEDs is one thing, but lighting up numbers and letters on a seven-segment display is much more engaging.

If your development board doesn't meet all these requirements, don't worry: you can still work through this book's projects with a few adjustments. For example, if you'd prefer to work with a board built around a different kind of FPGA, you can. As we'll discuss in later chapters, there are advanced features that vary from one FPGA to another, but the code for this book's projects is general enough that it should work on any modern FPGA. That's part of the beauty of Verilog and VHDL: they're FPGA-agnostic.

Do be aware that if you aren't working with an iCE40 FPGA, however, you'll need to use a different set of software tools than the ones discussed in this chapter. Each FPGA company provides its own tools specifically aimed at its FPGAs. For example, AMD (Xilinx) has Vivado, and Intel (Altera) has Quartus. If your board has an FPGA from one of these companies, look online for resources about using the appropriate software.

If you don't have all the necessary peripherals for the projects in this book, you have a few options. First, you can modify the projects' Verilog or VHDL code to use fewer LEDs and switches. This will work in most cases, although the memory game project in Chapter 8 will be less satisfying the fewer LEDs and switches you use.

Alternatively, many FPGA development boards, including some of the boards discussed in Appendix A, have connection points for wiring up your own peripherals. In particular, look for a development board with a Pmod (peripheral module) connector. Pmod is a standard connector made famous by Digilent for attaching accessory boards with extra peripherals—not just the ones used in this book, but also devices like temperature sensors, accelerometers, audio jacks, microSD cards, and more. If you've ever worked with Arduino Shields, it's the same concept. If your board has a Pmod connector, that will greatly expand the range of projects you can work on with your FPGA.

Setting Up Your Development Environment

To use the iCE40 FPGA on your development board, you'll need to install two software tools on your computer: iCEcube2 and Diamond Programmer. These free tools from Lattice Semiconductor are designed specifically for working with iCE40 FPGAs. This section walks you through the process of setting them up. If you're on Windows, you'll have the easiest time, since the tools are designed for the Windows operating system. For Linux or macOS users, I recommend creating a Windows virtual machine on your computer, then running the Lattice tools in that. There are many tutorials online for setting up a Windows virtual machine using VirtualBox or a similar product.

iCEcube2

iCEcube2 is Lattice's free integrated development environment (IDE) for turning the VHDL or Verilog code you write on your computer into a file that the FPGA can be programmed with. It's much easier to use than other IDEs like Vivado, Quartus, or even Lattice Diamond (not to be confused with Diamond Programmer), Lattice's tool for working with more sophisticated FPGAs. Their compatibility with iCEcube2 is part of what makes iCE40 FPGAs an especially good choice for beginners. Those other programs are all several gigabytes in size and extremely complicated. They have many bells and whistles, most of which you won't need when you're getting started. By contrast, iCEcube2 is more streamlined, making it a more straightforward tool for learning about FPGAs.

To download and install iCEcube2, follow these steps:

1. Visit *https://latticesemi.com/icecube2* or search the internet for "iCEcube2 download."

2. Find the download link for the latest Windows version of iCEcube2, whether you're running Windows natively or in a virtual machine. If you're a Linux user, you may be tempted to download the Linux version instead, but I wouldn't recommend it. That version is buggy; you might have success, or you might not.

3. When you click the download link, you'll be asked to create an account on the Lattice website. You must create an account to get a license for this tool. Make sure to use a real email address, as they'll email you the free license. Once you create an account, you should be able to download the software.

4. You'll need to request a license, which is free for hobbyists and similar users but can take a couple days to arrive in an email. Follow the steps in the "Licensing" section of the iCEcube2 page to start the process.

5. You'll need your computer's MAC address to obtain the license. To find it on Windows, open a command prompt by clicking the Start button and searching for "cmd." Then enter **ipconfig /all** at the command line. You should see something like this:

```
C:\> ipconfig /all
--snip--
Ethernet adapter Local Area Connection:

    Connection-specific DNS Suffix  . :
    Description . . . . . . . . . . . : Intel(R) Ethernet Connection I217-V
    Physical Address. . . . . . . . . : 38-D3-21-F5-A3-09
    DHCP Enabled. . . . . . . . . . . : Yes
    Autoconfiguration Enabled . . . . : Yes
```

6. Your MAC address is the 12-digit hexadecimal number next to Physical Address. Make a note of this MAC address as it will be required for the license request process.

7. Launch the iCEcube2 installer once it finishes downloading and point it to your license file.

NOTE *If you already installed iCEcube2 before obtaining the license, you can use the program* LicenseSetup.exe *in the same folder where you installed the tool to point it at your license file.*

When it's done installing, launch iCEcube2. The main window will look something like Figure 2-1.

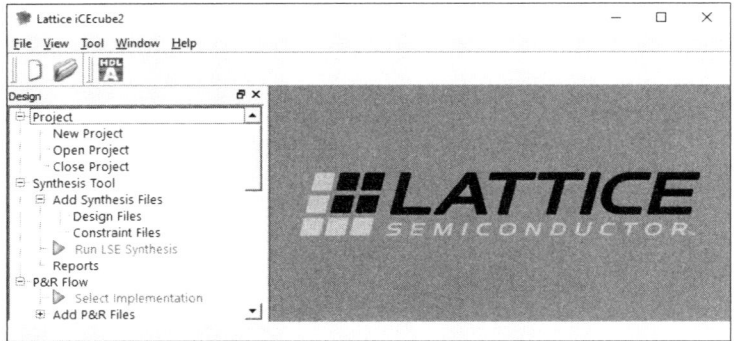

Figure 2-1: The iCEcube2 main window

Click around to get a feel for the program. We'll explore it in more detail later in this chapter with a project that will take you through the entire FPGA build process.

Diamond Programmer

Diamond Programmer is a free standalone programming tool from Lattice that takes the output of iCEcube2 and uses it to program your FPGA via your development board's USB connection. More sophisticated software tools like Vivado and Quartus have a built-in programmer, so you don't need to download a separate program. It's unfortunate that iCEcube2 doesn't have one built in, but such is the life of an iCE40 FPGA designer! Here's how to install Diamond Programmer:

1. Go to *https://www.latticesemi.com/diamond* or search the internet for "lattice diamond software" to locate the download page.
2. The Software Downloads section has many download links to choose from. Find and click the link for the latest version of Programmer Standalone 64-bit for Windows.

WARNING *Be sure to download Programmer Standalone and not Diamond or the Programmer Standalone Encryption Pack. The latter isn't needed.*

3. Diamond Programmer doesn't require a license, so simply run the installer once it downloads.

You're now ready to dive into your first FPGA project, where you'll learn how to work with these tools and program your FPGA.

Project #1: Wiring Switches to LEDs

In this project, you'll get familiar with the build process by creating a simple FPGA design: when you press one of the push-button switches on your FPGA development board, one of the LEDs should light up. The project assumes that you have four switches and four LEDs, so you'll design and program your FPGA to wire up each switch to one of the LEDs. (As mentioned earlier, you can adapt the project to use fewer switches and LEDs if needed.) Figure 2-2 shows a diagram of what we want to do.

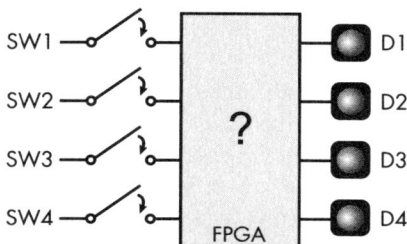

Figure 2-2: The Project #1 block diagram

On the left we have the board's four switches, labeled SW1 through SW4. By default, these will be open (not connected), meaning the corresponding input pin of the FPGA will have a low voltage when the switch isn't pressed, due to an onboard pull-down resistor. When you press a switch down, the FPGA will see a high voltage present at the input pin connected to that switch. On the output side we have four LEDs, labeled D1 through D4. We want to create an FPGA that will connect the switches and LEDs such that, for example, when the user presses SW1, the D1 LED illuminates. We'll literally be creating a physical wired connection between the SW1 input and the D1 output using our FPGA. In other words, with FPGAs you're programming at such a low level that you're creating wires between pins, throughout your device.

To implement this project, we'll go through four main steps. These steps, summarized in Figure 2-3, form the main phases of the FPGA build process.

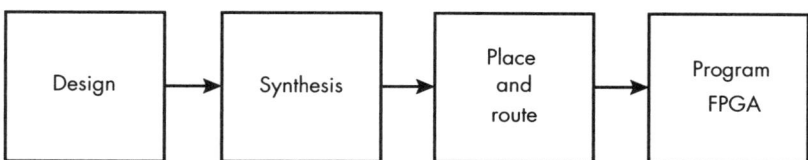

Figure 2-3: The FPGA build process

You'll familiarize yourself with the steps at a high level through this project. Then you'll expand your knowledge of each step throughout the book. The four steps are:

1. **Design.** In this step, you write the Verilog or VHDL code that describes how the FPGA will function. You might also write tests to ensure that your code will work as you intend, a concept we'll discuss in Chapter 5.

2. **Synthesis.** The synthesis process is what turns your code into low-level components that perform the actual functionality on your FPGA. It's similar to how a compiler in a programming language like C turns your C code into Assembly instructions. In this book, we'll use iCEcube2 as a synthesis tool.

3. **Place and route.** This process will take your synthesized design and map it to the physical layout of your specific FPGA. It will wire up (route) the connections between components, including connecting the input and output pins to the inner FPGA components. Creating links between pins and signals in your code is one of the purposes of the *physical constraints file.* You'll see how to write a constraint file in this project. iCEcube2 handles the place and route step at the same time that it handles synthesis.

4. **Programming.** This is where you take the output of the previous steps and load it onto your physical FPGA. The programming file literally creates wired connections between pins and FPGA components, and within the FPGA itself. This project will simply create wires between pins, but in future projects we'll use other FPGA components as well. The programming step happens within Diamond Programmer.

All the projects in this book will follow this same basic process. As you work on later projects, refer back to this section if you need a refresher on using iCEcube2 and Diamond Programmer.

Writing the Code

Let's design an FPGA that links the switch inputs to the LED outputs using Verilog or VHDL. Hopefully by this point you've chosen which language you want to learn; I suggest focusing on just one for now, but you can always pick up the other later. All of this book's code examples are shown in both languages, so you can compare and contrast the code as well.

I've had success writing FPGA code with Visual Studio Code (VS Code), a free tool from Microsoft. You can download extensions that will enable Verilog or VHDL syntax highlighting and other useful features, like the ability to tie to GitHub repositories directly from the code editor. You could also write your code directly in iCEcube2, but I wouldn't recommend it, as it doesn't have syntax highlighting.

Whatever tool you choose, enter the following Verilog or VHDL code and save it on your computer. Take note of the filename and location, as you'll need it later. All of the code in this book is also available in the book's GitHub repository, *https://github.com/nandland/getting-started-with-fpgas.*

Verilog

```
❶ module Switches_To_LEDs
❷  (input  i_Switch_1,
    input  i_Switch_2,
    input  i_Switch_3,
    input  i_Switch_4,
❸ output o_LED_1,
   output o_LED_2,
   output o_LED_3,
   output o_LED_4);

❹ assign o_LED_1 = i_Switch_1;
   assign o_LED_2 = i_Switch_2;
   assign o_LED_3 = i_Switch_3;
   assign o_LED_4 = i_Switch_4;

endmodule
```

VHDL

```
library ieee;
use ieee.std_logic_1164.all;

❶ entity Switches_To_LEDs is
   port (
❷ i_Switch_1 : in std_logic;
   i_Switch_2 : in std_logic;
   i_Switch_3 : in std_logic;
   i_Switch_4 : in std_logic;
❸ o_LED_1    : out std_logic;
   o_LED_2    : out std_logic;
   o_LED_3    : out std_logic;
   o_LED_4    : out std_logic);
end entity Switches_To_LEDs;
architecture RTL of Switches_To_LEDs is
begin

❹ o_LED_1 <= i_Switch_1;
   o_LED_2 <= i_Switch_2;
   o_LED_3 <= i_Switch_3;
   o_LED_4 <= i_Switch_4;

end RTL;
```

Let's consider broadly how this code is structured, since all our projects will follow this same general format. The design for an FPGA is encapsulated inside one or more *modules* (in Verilog) or *entities* (in VHDL). These modules/entities define the interface to a block of code. The interface has signals, which can be inputs or outputs. At the highest level of your FPGA, these signals will connect to physical pins on your device, thereby creating the interfaces to other components, such as switches and LEDs.

To create a module in Verilog, you use the module keyword and provide a descriptive name—in this case, Switches_To_LEDs ❶. Inside the module, the first thing you do is declare all the input ❷ and output ❸ signals, enclosed in a set of parentheses. Then comes the code for what you want the module to actually do, which we'll discuss in detail momentarily, followed by the `endmodule` keyword.

Looking at the VHDL version, the first thing you might notice is that it's a bit longer than the Verilog version. This is typical; VHDL generally takes more typing to accomplish the same task compared to Verilog. Some of the extra length comes at the very beginning of the listing, where we specify which VHDL library and package we'll be using. In this case, we use the `std_logic_1164` package from the `ieee` library. We need this to get access to the `std_logic` data type, which is commonly used to represent binary values (0, 1) within your FPGA. Get used to including this library and package. You'll need it for every VHDL design you create.

Whereas in Verilog you declare the inputs and outputs and code the actual logic of the module as part of the same code block, in VHDL you do this with two separate code blocks. This is another reason why the VHDL version is longer. First, you use the `entity` keyword to declare the VHDL entity ❶, giving it a name and specifying its inputs ❷ and outputs ❸. Then, in a separate code block, you use the `architecture` keyword to declare the *architecture* of the entity, which is the code that defines the entity's functionality. You'll almost always have a single entity/architecture pair in a VHDL file, with the entity describing the input/output interface and the architecture describing the functionality.

Now that we've covered the structure of the code, let's look at the specifics. In both the Verilog and VHDL versions, we define the four input signals ❷ corresponding to the four switches: `i_Switch_1`, `i_Switch_2`, `i_Switch_3`, and `i_Switch_4`. In Verilog, these inputs will be defined as 1 bit wide (a single 0 or 1) by default, whereas in VHDL we explicitly define them as `std_logic`, which is a 1-bit-wide data type. We similarly define the four outputs, `o_LED_1`, `o_LED_2`, `o_LED_3`, and `o_LED_4`, for the four LEDs ❸. Notice that I like to precede my input signal names with `i_` and my output signal names with `o_`. This helps me to keep track of which direction each signal is going in.

NOTE *You can define your inputs and outputs in any order, but it's customary to put inputs first.*

Finally, we define the logic of the design—the code that actually does the work—by *assigning* the inputs to the outputs ❹. For example, we take the value on input `i_Switch_1` and assign it to the output `o_LED_1`. When the FPGA is built, this will create a physical wire between these two pins. In Verilog we use the `assign` keyword, which requires the `=` for the actual signal assignment. In VHDL, we can just use the `<=` assignment to create the wire between the input and the output.

Creating a New iCEcube2 Project

Once you have the coding done, it's time to bring the design into iCEcube2 so you can build it. Open iCEcube2 and select **File ▶ New Project**. You'll be taken to a window asking for information about your FPGA board, as shown in Figure 2-4. Let's review the settings in this window.

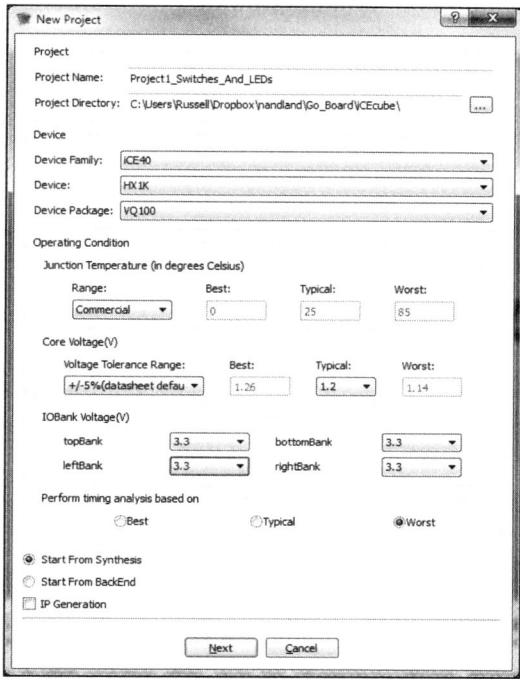

Figure 2-4: The iCEcube2 New Project window

For the Project Name, give your project whatever name you like, and for the Project Directory, choose where on your computer you want it saved. Next, you need to tell the tool which FPGA you're using. It needs to know how many resources the FPGA has, which pins go where, and everything about how it works to properly turn your code into something compatible with your specific device. To do this, select **iCE40** from the Device Family drop-down, then choose your FPGA's specific device name and package from the Device and Device Package drop-downs. For example, if you were using the Nandland Go Board (one of the boards discussed in Appendix A), you would choose HX1K for the device and VQ100 for the package, then select 3.3 from the topBank, leftBank, bottomBank, and rightBank drop-downs. This tells the tool that all the pins on the device operate at 3.3 volts. Everything else in the window can remain at the default settings. Click **Next** when you're done.

NOTE   *You'll use these same settings for every single project, so you can refer back to this section each time you create a new project.*

You'll be taken to another dialog that prompts you to add the Verilog or VHDL source file that you created previously. Go ahead and add your file, or click **Finish** to skip this step for now. If you choose to skip adding your file from the dialog, you can do so later by expanding the Synthesis Tool menu on the left side of the main iCEcube2 project window, right-clicking Design Files, and selecting Add Files, as shown in Figure 2-5.

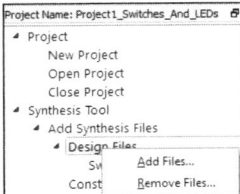

Figure 2-5: Adding Verilog or VHDL source files to your project

This Design Files menu also makes it possible to add additional files to an existing project after it's been created, or to remove and replace files that you've added previously.

Adding Pin Constraints

The next step in the build process is to add the pin constraints to your project. These constraints, which you declare in a *.pcf (physical constraints file)* file (sometimes referred to as pin constraint file), tell the tool which signals in your Verilog or VHDL code will be connected to which physical pins on your FPGA. This information is critical to the place and route stage of the build process, when the output of the synthesis process is mapped to the physical resources on your FPGA. The tool needs to know which pins are connected to the switches and LEDs so all the wires in the design can be routed to where they need to go.

Each FPGA manufacturer has its own keywords for writing constraints. To declare a pin constraint for Lattice's iCEcube2, you use the set_io keyword, followed by the name of one of the signals in your design, followed by the corresponding pin number on the FPGA itself. Here's an example of what the physical constraint file should look like for this project, but keep in mind that the actual pin numbers will vary depending on your development board. As an example, these pin numbers would work for the Nandland Go Board:

```
# LED pins:
❶ set_io o_LED_1 56
set_io o_LED_2 57
set_io o_LED_3 59
set_io o_LED_4 60

# Push-button switches:
set_io i_Switch_1 53
set_io i_Switch_2 51
set_io i_Switch_3 54
set_io i_Switch_4 52
```

Each line maps one of the signals in our code to one of the pins on the FPGA. For example, we set the Verilog/VHDL signal o_LED_1 to be connected to pin 56 on the FPGA ❶. The signal names you use in the physical constraint file must match the signal names in your Verilog/VHDL code exactly. If the names don't match, the tool won't know which signal goes to which physical pin on the device.

Notice that comments in the physical constraint file are preceded with a # symbol—an octothorpe, pound sign, or hashtag, depending on your age.

When setting your pin constraints, you'll need to look at the reference schematic for your FPGA development board. The schematic contains the wiring diagram for the circuit board. It tells you which pin of the FPGA is connected to which LED, button, connector pin, or other device. Learning how to read this basic schematic information is a critical skill for an FPGA designer, as setting pin constraints is a common task.

To add the physical constraint file to your project, find the P&R Flow section in the menu on the left side of the iCEcube2 project window, expand **Add P&R Files**, and right-click **Constraint Files**. Then click **Add Files** and select your *.pcf* file. Once you do this, you'll see the file listed under Constraint Files.

Forgetting to add a physical constraint file is a common mistake when working with FPGAs. If you don't add one, the tool won't warn you about it. Instead, they'll just connect the signals in your code to randomly chosen pins on your device. This will almost certainly be wrong, and your design won't work as you expect.

Running the Build

You're now ready to run the build in iCEcube2. To do this, simply click **Tool ▸ Run All**. This will execute both the synthesis and place and route processes, creating the FPGA image file that you'll use to program the FPGA. iCEcube2 generates a report for each of these steps, visible under the Reports section. You're welcome to explore these reports to see what type of information they contain; we'll dive into the details in future chapters.

Connecting Your Development Board

You now need to connect your board to your computer to program the FPGA. Take a minute to make sure this connection works and that your computer recognizes the device. With the board unplugged, open up Device Manager in Windows and expand the Ports (COM & LPT) section. Now go ahead and plug in the board via USB. You should see two devices labeled "USB Serial Port (COM*X*)" pop up, as shown in Figure 2-6. The specific COM port index numbers don't matter. If this works for you, then your board is connected to your computer and you're ready to go.

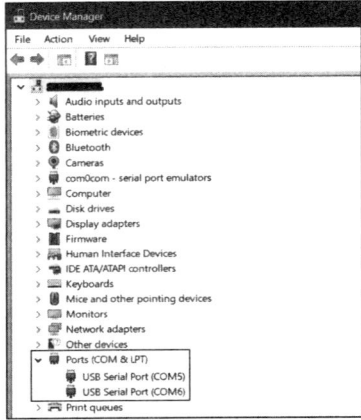

Figure 2-6: Viewing the board connection in Device Manager

If you don't see the USB serial ports in Device Manager, there are a few troubleshooting techniques to try. First, check if your board has a power LED for indicating when it's on. If it does, but that LED isn't illuminated, you don't have power, so check that the USB cable is firmly plugged into the board and into your computer. If the LED *is* illuminated, then the next most likely issue is the USB cable itself. Some Micro-USB cables are "charge only," meaning they don't have the wires that allow for data transfer. Get another cable that you know works to transfer data to and from a computer.

Programming the FPGA

The final step in the process is to program your design to your FPGA using Diamond Programmer (from the Programmer Standalone download earlier). An FPGA development board typically features an integrated circuit that turns its USB connection into the SPI interface, which Diamond Programmer uses to program a flash memory chip installed on the board. Once that's done, the FPGA will boot up from the flash, and you'll see the fruits of your labor!

With your board connected, open up Diamond Programmer to get started. You'll be greeted with the dialog shown in Figure 2-7. Click **OK** to create a new project.

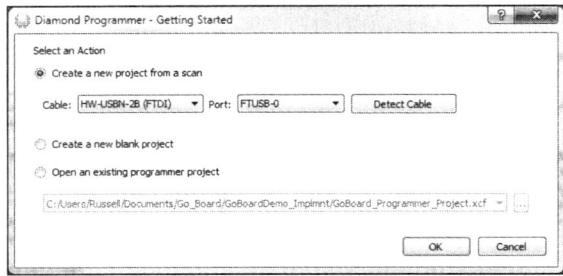

Figure 2-7: The Diamond Programmer dialog

Once you click OK, the tool will try to scan the board to automatically identify which FPGA is connected. It will fail. That's fine; we can manually tell Diamond Programmer which FPGA to target from the next screen, which is shown in Figure 2-8.

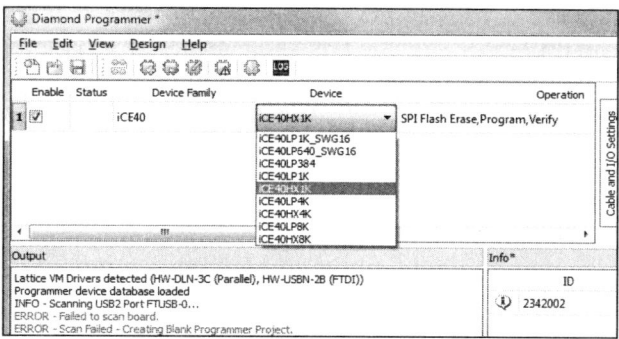

Figure 2-8: The Diamond Programmer device selection screen

Set the Device Family to **iCE40** and choose your specific FPGA from the Device drop-down, as shown in Figure 2-8. Next, double-click the field under Operation. You'll be greeted with a new window, shown in Figure 2-9. Note that you may need to change the access mode to SPI Flash Programming to see the contents shown here.

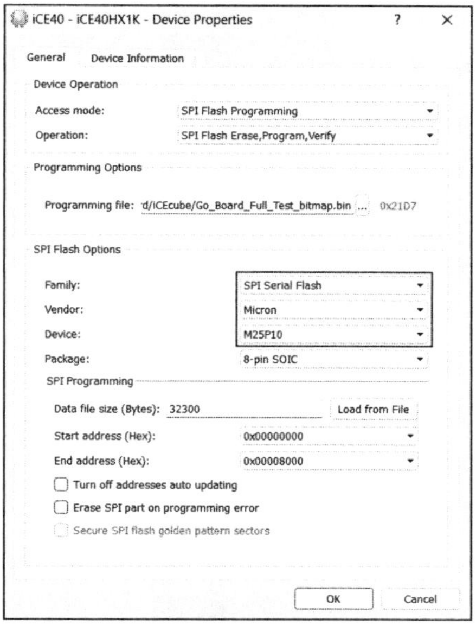

Figure 2-9: The Diamond Programmer
Device Properties window

This window lets you tell Diamond Programmer how to program your FPGA. In the Device Operation section, set the access mode to **SPI Flash**

Programming. For the SPI Flash Options section, you'll have to consult the programming guide for your development board to determine which SPI flash device is used. For the Go Board, for example, you'd set the family to SPI Serial Flash, the vendor to Micron, and the device to M25P10, as shown in Figure 2-9.

Finally, in the Programming Options section, click the three dots next to the Programming File box and choose the *.bin* file to program to the FPGA. This is the file you generated using iCEcube2, located in the */<Project_Name>_Implmnt/sbt/outputs/bitmap/* subdirectory inside the directory where you saved your iCEcube2 project. Leave all the other settings on their defaults, and click **OK** to close this window. Now you're ready to program.

Open the **Design** menu and select **Program**. If everything was done correctly, you should see INFO – Operation: successful after a few seconds. This means that your SPI flash has been programmed and your FPGA is running! Try pushing each switch on your board. You should see the corresponding LED light up when the button is held down. Congratulations, you've built your first FPGA project!

NOTE *I recommend saving your Diamond Programmer project so you can reuse the settings for the other projects in the book. All you'll have to do is select a different .bin file to program to the FPGA.*

If the programming fails, you might get a CHECK_ID error like this:

```
ERROR – Programming failed.
ERROR – Function:CHECK_ID
Data Expected: h10   Actual: hFF
ERROR – Operation: unsuccessful.
```

If you see this error, go to the Cable Settings section in the right pane of Diamond Programmer and change your port from FTUSB-0 to **FTUSB-1**, as shown in Figure 2-10.

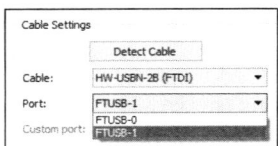

Figure 2-10: Troubleshooting a CHECK_ID error

Once you make the change, try to program your device again. This time it should work.

Summary

In this chapter, you created an FPGA development environment and learned how to work with a development board. Through your first project, you learned about the main steps in the FPGA development process: design, where you write code for the FPGA using Verilog or VHDL; synthesis, where you translate that code into FPGA components; place and route, where you map the synthesized design to the resources on your specific FPGA; and programming, where the design is physically transferred to the FPGA. We'll explore these concepts in more detail later in the book, but as you work through other projects, remember that you can refer back to this chapter if you need a refresher on the basics of using your FPGA tools.

3

BOOLEAN ALGEBRA AND THE LOOK-UP TABLE

Boolean algebra is a field of mathematics and logic essential to understanding how to work with devices like FPGAs. In Boolean algebra, the input and output values are all true or false, which we can equate with 1s and 0s, or high and low voltages. Instead of operations like multiplication and division, Boolean algebra has operations such as AND, OR, and NOT. Each of these operations takes in some number of 0s and 1s as input, evaluates them, and produces a 0 or 1 as an output. Algebra class would have been much simpler if it had been about Boolean algebra!

You may have encountered Boolean operations in other programming languages, such as C or Python. For example, you might want your program to write to a file only if the user chooses to do so *and* the filename is valid. Similarly, inside an FPGA, you'll often want to check multiple inputs to determine the state of an output. Let's say you want to turn on an LED when either of two switches is pressed. An FPGA can make this happen using an OR operation: if either one switch *or* the other (or both) provides a 1 as an input, the FPGA provides a 1 as an output to the LED, and the LED lights up.

Boolean algebra makes tasks like this possible. But more importantly, Boolean algebra describes *all* the underlying operations on data in your FPGA. String together enough Boolean operations and you can do math, store data, and more. You can do a surprising amount by manipulating 1s and 0s.

In this chapter, we'll explore how to represent simple Boolean operations with logic gates, and we'll see how these gates can be combined into more complicated Boolean equations. Then we'll explore how FPGAs actually perform logic operations by combining the functionality of different logic gates into a single device called a look-up table. As you'll see, look-up tables are one of the most important components in an FPGA.

Logic Gates and Their Truth Tables

When designing FPGAs, we represent simple Boolean operations with *logic gates*, devices that take in electrical signals as inputs, perform a Boolean operation on them, and produce the appropriate electrical signal as an output. There are different logic gates corresponding to all the different Boolean operations, such as AND, OR, NOT, XOR, and NAND. Each of these logic gates can be described with a *truth table*, a table that lists all the possible input combinations for a Boolean algebra equation and shows the corresponding outputs.

We'll discuss some common logic gates and examine their truth tables next. But first, it's important to understand what the 1s and 0s in the truth tables we'll be looking at actually mean. Inside an FPGA, digital data is represented by voltages: 0 volts for a 0, and some higher-than-zero voltage, called the *core voltage*, for a 1. The core voltage depends on the specific FPGA, but often is around 0.8 to 1.2 volts. When we talk about a signal being *high*, we mean that the signal is at the core voltage and represents a data value of 1. Likewise, a *low* signal is at 0 volts and represents a data value of 0. With this in mind, let's look at some logic gates.

AND Gates

An *AND gate* is a logic gate whose output is high when all its inputs are high. We'll use the example of a two-input AND gate, but AND gates can have any number of inputs. For a two-input AND gate, the output is high when input A *and* input B are both high, hence the name AND gate. Table 3-1 shows the truth table for this AND gate. Notice that the output is a 1 only when both inputs are a 1.

Table 3-1: Truth Table for a Two-Input AND Gate

Input A	Input B	Output Q
0	0	0
0	1	0
1	0	0
1	1	1

In a truth table, the rows are usually arranged in increasing decimal order, based on the inputs. In the case of the AND truth table, the first row shows when input A = 0 and input B = 0, which is represented as b00, which means 00 in binary, or 0 in decimal. Next comes b01 (decimal 1), then b10 (decimal 2), then b11 (decimal 3). If the AND gate had additional inputs, then there would be more rows in our truth table that we would have to fill out. In the case of a three-input AND gate, for example, there would be eight rows, going from b000 to b111, or 0 to 7 in decimal.

NOTE *The output of a logic gate is denoted with a Q. This convention comes from the English mathematician Alan Turing, who used the letter Q to denote states in his famous Turing machines. The Q stood for* quanta, *which is a discrete state (such as 0 or 1), rather than something that can have a continuous range of values.*

Each logic gate has a distinctive symbol for use in schematics. A two-input AND gate is drawn as shown in Figure 3-1. The symbol depicts the inputs A and B going into the gate on the left, and the output Q emerging on the right.

Figure 3-1: The AND gate symbol

As we continue our exploration of logic gates, most of the gates we'll look at will have two inputs and one output. As with AND gates, it's possible that these other types of gates could have additional inputs, but for simplicity we'll stick to the two-input versions. (The exception is the NOT gate, which can only have one input and one output.) For brevity, I'll omit the words *two-input* from this point forward when referring to a given logic gate.

OR Gates

An *OR gate* (Figure 3-2) is a logic gate whose output is high when either of the inputs is high; that is, when either input A *or* input B is high.

*Figure 3-2: The OR
gate symbol*

Table 3-2 shows the truth table for an OR gate.

Table 3-2: Truth Table for an OR Gate

Input A	Input B	Output Q
0	0	0
0	1	1
1	0	1
1	1	1

Notice that when both inputs are high, the OR gate's output is high as well. All that matters to an OR gate is that at least one of the inputs is high, which is also the case when both inputs are high.

NOT Gates

A *NOT gate* (Figure 3-3) has a single input and a single output. This kind of gate simply inverts the input (the output is *not* the input), so it's also known as an inverter.

*Figure 3-3: The NOT
gate symbol*

Notice the bubble at the tip of the triangle in the NOT gate symbol, which indicates inversion. It also appears in the NAND gate, which we'll look at later, and can even appear on some inputs. The truth table for a NOT gate is shown in Table 3-3.

Table 3-3: Truth Table for a
NOT Gate

Input A	Output Q
0	1
1	0

As the truth table indicates, whatever the input value to the gate is, the output is the opposite.

XOR Gates

The output of an *XOR gate* (pronounced "ex-or," short for *exclusive or*) is high when either of the inputs is high, but not both. In other words, the gate checks for exclusively one or the other input being high. The symbol for an XOR gate is shown in Figure 3-4.

Figure 3-4: The XOR gate symbol

The symbol looks like that of an OR gate, but the extra line on the left side of the gate sets it apart. Table 3-4 shows the XOR gate's truth table.

Table 3-4: Truth Table for an XOR Gate

Input A	Input B	Output Q
0	0	0
0	1	1
1	0	1
1	1	0

Though this type of gate might not seem particularly useful at first blush, it comes up more often than you might expect. For example, XOR gates are used for generating a *cyclic redundancy check (CRC)*, a way to validate data to verify the integrity of transmitted information.

NAND Gates

A *NAND gate* (short for *not and*) has the opposite output of an AND gate. You can infer this from the NAND gate's schematic symbol, shown in Figure 3-5: it looks exactly like an AND gate, except with a bubble on the output to indicate an inversion.

Figure 3-5: The NAND gate symbol

The output of the NAND gate is thus the same as an AND gate, but inverted. If both input A and input B are high, output Q will be low. In all other cases, output Q will be high. This is shown in the truth table in Table 3-5.

Table 3-5: Truth Table for a NAND Gate

Input A	Input B	Output Q
0	0	1
0	1	1
1	0	1
1	1	0

NAND gates are commonly used in USB flash drives, solid state drives (SSDs), and other types of data storage devices. They also inspired the name of my website, *https://nandland.com*.

Other Gates

We've explored the most common types of logic gates here to give you an idea of how they work, but this isn't an exhaustive list. There are other types as well, such as NOR (short for *not or*) and XNOR (*exclusive not or*) gates. Additionally, as mentioned previously, though we focused on the two-input versions here, all of these gates (with the exception of NOT) can have more than two inputs. This section was just intended to get you comfortable with the standard logic operations from Boolean algebra. Next, we'll explore how these operations can be combined to make more complicated expressions.

Combining Gates with Boolean Algebra

You've seen how individual logic gates work. However, often you'll want to write code that's more complex than just a single logic operation. The good news is that you can chain together multiple logic gates to represent more elaborate Boolean equations, and use Boolean algebra to determine the outcome.

In Boolean algebra, each logic operation has its own symbol. One common set of symbols is shown in Table 3-6. For example, * represents an AND operation, and + represents an OR operation. These symbols make it easier to write more elaborate Boolean algebraic equations.

Table 3-6: Boolean Algebra Symbols

Symbol	Meaning
*	AND
+	OR
'	NOT
^	XOR

Boolean algebra also has its own *order of operations*. To solve a Boolean equation, first you evaluate NOTs, then ANDs, and finally ORs. As in conventional algebra, you can use parentheses to bypass the order of operations; anything in parentheses will be evaluated first.

You now know everything you need to write and evaluate Boolean equations with more than one logic operation, such as $Q = A * B + A'$. In plain language, you'd read this as "The output Q equals A *and* B *or not* A." Table 3-7 shows the truth table for this equation.

Table 3-7: Truth Table for A * B + A'

Input A	Input B	Output Q
0	0	1
0	1	1
1	0	0
1	1	1

Figure 3-6 shows the circuit equivalent of this equation, created by combining logic gates.

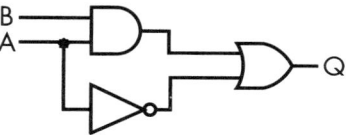

*Figure 3-6: The circuit diagram for A * B + A'*

As you can see, we still have only two inputs, but because those inputs go through three different logic operations, the possible outputs of our truth table are more interesting than they were for individual logic gates. Let's consider what happens with this equation when both inputs are 0, the first row of our truth table. The equation has no parentheses, so first we look at NOT A, which evaluates to 1. Then we perform the AND operation of A and B, which evaluates to 0. Finally, we OR the results of both of those expressions, giving us an output of 1. Considering the other possible inputs, you should see that any time A is 0, or any time A and B are both 1, the output Q will be 1. Otherwise, the output will be 0.

While this example featured two inputs, it's possible to have Boolean equations with any number of inputs. Each input increases the number of rows in the truth table by a factor of 2: for one input there are two truth table rows, for two inputs there are four rows, for three inputs there are eight rows, and so on. In mathematical terms, for n inputs, there are 2^n truth table rows.

To demonstrate, let's consider an example equation with three inputs: $Q = A + (C * B')$. Note that the parentheses indicate that the operation C AND NOT B occurs prior to the OR operation. In fact, that follows the regular Boolean algebra order of operations, but the parentheses make the equation a little easier to read. The truth table with three inputs is shown in Table 3-8.

Table 3-8: Truth Table for A + (C * B')

Input A	Input B	Input C	Output Q
0	0	0	0
0	0	1	1
0	1	0	0
0	1	1	0
1	0	0	1
1	0	1	1
1	1	0	1
1	1	1	1

The corresponding circuit is shown in Figure 3-7.

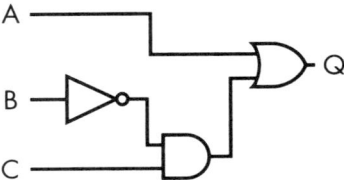

*Figure 3-7: The circuit diagram for A + (C * B')*

To generate this truth table, first we should perform the operation inside the parentheses. This is C AND NOT B. Within the parentheses, the highest precedence is the inversion applied to B, which is then ANDed with C. In all, the part of the equation in parentheses evaluates to high when C is high and B is low, and since the remainder of the equation is an OR operation, we also know that the overall output will be high when C is high and B is low. This case occurs on the second line of the truth table. It also occurs on the sixth line of the truth table, so we can fill those in with a 1. Finally, consider the A on the other side of the OR operation. When it's high, as in the last four lines of the truth table, the output will be high. We can fill in the remaining lines with a 0 to complete the truth table.

Combining logic operations to perform more complicated functionality is common throughout programming. In FPGAs, this same capability is possible by chaining together simple logic gate functions.

The Look-Up Table

So far we've been learning about individual logic gates, but it might surprise you to find out that these logic gates don't physically exist inside an FPGA. There isn't a bank of AND gates and OR gates that you can just pull from and wire together to create your Boolean algebra logic. Instead, there's something much better: *look-up tables (LUTs)*. These are devices that can be programmed to perform any Boolean algebra equation you can think of, regardless of the specific logic gate(s) involved. If you need an AND gate, a LUT can do it. If you need an XOR gate, a LUT can do that too. A single LUT can also evaluate an equation involving multiple logic gates, like the ones we considered in the previous section. Any truth table you can think of, a LUT can produce. This is the power of the look-up table.

NOTE *Early programmable logic devices like Programmable Array Logic (PAL) did actually have banks of AND and OR gates. With FPGAs, these have been superseded by the more capable LUTs.*

LUTs are classified by the number of inputs they can accept. For example, there are two-, three-, four-, five-, and even six-input LUTs on the newest FPGAs. Most LUTs produce a single output. Figure 3-8 shows what a three-input LUT (often referred to as LUT-3) looks like.

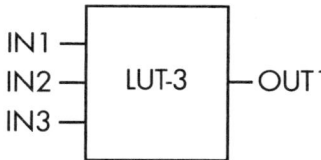

Figure 3-8: A three-input LUT

This LUT is a blank slate that can be programmed to perform any Boolean algebra operation with three inputs and one output. For example, look back at the circuit drawn in Figure 3-7 to represent the Boolean equation $Q = A + (C * B')$. Drawing the circuit for that equation required three logic gates—a NOT gate, an AND gate, and an OR gate—but we can replace those three gates with our single three-input LUT. The same LUT could also be programmed to represent the equation $Q = (A + B + C)'$, or $Q = (A + B)' * C$.

What happens if we have a Boolean algebra equation with more than three inputs? That's just fine, as LUTs can be chained together to perform very long sequences of logic. In fact, the typical FPGA contains hundreds or even thousands of LUTs, all ready to be programmed to carry out whatever logic operations you need. This is why look-up tables are one of the two most important components to understand inside of an FPGA: they perform the logical operations of your code. The other key component is the flip-flop, which we'll talk about in the next chapter.

Although we've been drawing truth tables and logic gate diagrams here, in the real world you'll rarely define FPGA operations this way. Instead, you'll write code. Often, the code you write is at a higher level than individual logic gates: you might write code to compare two numbers, or increment a counter, or check if a condition is true, and the synthesis tools then break down that code into the necessary Boolean logic operations and assign those operations to LUTs. However, the purpose of this book is to teach you how FPGAs work so you understand them, and at a fundamental level, FPGAs work by performing Boolean algebra. Once you know how

FPGAs work, you'll be able to use Verilog or VHDL with a deeper under-standing of what you're creating with your code. This will help you create efficient and reliable FPGA designs.

LOGIC MINIMIZATION

You'll often hear about *logic minimization techniques* such as De Morgan's law, Karnaugh maps, and the Quine–McCluskey algorithm in connection to Boolean algebra. These are mathematical tricks to simplify Boolean algebra equations to take up fewer computing resources. I feel that knowing the ins and outs of these techniques isn't necessary to begin learning about FPGAs, however. Yes, LUTs are limited and should be used optimally, but there are entire software tools responsible for minimizing logic resource usage for you. Specifically, synthesis tools, which we'll discuss in more detail in Chapter 7, perform this task so you don't have to. I've spent weeks of my life learning about Karnaugh maps and performing Quine–McCluskey by hand, and I can tell you that I've never needed to use this knowledge as a professional FPGA engineer. All you need to do is write the VHDL or Verilog and let the software tools work their magic.

Project #2: Lighting an LED with Logic Gates

You're now ready to combine everything you've learned about Boolean logic and look-up tables in a real-world example on your FPGA development board. This project should illuminate an LED, but only when two switches are pushed at the same time. In other words, you're using your first LUT by implementing an AND gate. Figure 3-9 shows the block diagram for this project.

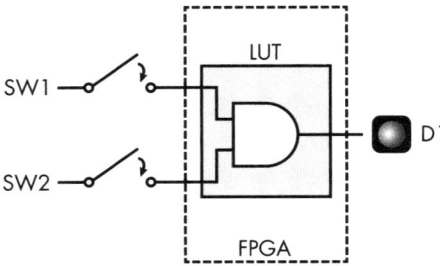

Figure 3-9: The Project #2 block diagram

This project turns the entire FPGA into one single AND gate. This might be overkill for a very capable FPGA, but it's an excellent way to visualize how a LUT works in the real world. Table 3-9 shows the truth table corresponding to the project.

Table 3-9: Truth Table for Project #2

SW1	SW2	D1
0	0	0
0	1	0
1	0	0
1	1	1

This truth table looks exactly like the one we created for the AND gate, but the column labels have been replaced to represent two switches and an LED on your development board. As expected, the output D1 is only high when inputs SW1 and SW2 are both high.

Writing the Code

Implementing an AND gate uses very few resources: three connections (two input and one output) and a single LUT. Let's look at the Verilog and VHDL to get that LED to light up:

Verilog
```
module And_Gate_Project
  (input i_Switch_1,
   input i_Switch_2,
   output o_LED_1);

❶ assign o_LED_1 = i_Switch_1 & i_Switch_2;

endmodule
```

VHDL
```
library ieee;
use ieee.std_logic_1164.all;

entity And_Gate_Project is
  port (
    i_Switch_1 : in std_logic;
    i_Switch_2 : in std_logic;
    o_LED_1    : out std_logic);
end entity And_Gate_Project;

architecture RTL of And_Gate_Project is
begin
❶ o_LED_1 <= i_Switch_1 and i_Switch_2;
end RTL;
```

We begin by defining our inputs as i_Switch_1 and i_Switch_2, and our output as o_LED_1. Then we continuously assign the output with the AND of the two inputs ❶. In Verilog the symbol for an AND operation is &, whereas in VHDL and is a reserved keyword.

Building and Programming the FPGA

You're now ready to run your Verilog or VHDL through the build process discussed in Chapter 2. The synthesis tool will generate a report outlining the resource utilization on your FPGA. Here's the most interesting part of the report:

```
--snip--
Resource Usage Report for And_Gate_Project

Mapping to part: ice40hx1kvq100
Cell usage:
SB_LUT4          1 use

❶ I/O ports: 3
  I/O primitives: 3
  SB_IO          3 uses

  I/O Register bits:               0
  Register bits not including I/Os:  0 (0%)
  Total load per clock:

  Mapping Summary:
❷ Total  LUTs: 1 (0%)
```

This report tells us that three I/O ports (input/output ports, or pins, meaning connections to the outside world) on the FPGA have been used to implement our circuit design ❶ and, most importantly, that we've used a single LUT ❷. That (0%) on the last line is indicating the resource utilization on the FPGA. On this particular FPGA there are over 1,000 LUTs available for usage, and we're only making use of 1 of them. Since the synthesis report is showing one LUT with 0 percent resource utilization, the tool must be doing some rounding down here (1 / 1,000 = 0.1).

Go ahead and program your development board, and notice that the LED only illuminates when the two switches are held down together. The LUT is working!

Feel free to change the code around to implement a different Boolean operation than AND. For example, you could create an OR gate or an XOR gate using the | or ^ symbols in Verilog, or the or or xor keywords in VHDL. You could also try stringing together several operations to make the LED light up based on whatever crazy Boolean algebra equation you can think of, or try adding in more switch inputs or more LED outputs to implement more complicated truth tables. You can check that the synthesis tools are really generating the correct LUTs based on your code by writing out your own truth table using the switches as inputs and the LED as the output, then testing all possible switch combinations to see if they work as expected.

Summary

In this chapter you've learned about one of the two most important components of an FPGA: the look-up table. You've seen how a LUT can implement any Boolean algebra equation with a given number of inputs, from simple logic gates like AND, OR, NOT, XOR, and NAND to more complex equations that combine these gates. In the next chapter, we'll focus on the other crucial FPGA component: the flip-flop.

4

STORING STATE WITH THE FLIP-FLOP

Alongside the look-up table, the other main component in an FPGA is the *flip-flop*. Flip-flops give FPGAs the ability to remember, or store, state. In this chapter, we'll explore how flip-flops work and learn why they're important to the functioning of FPGAs.

Flip-flops make up for a shortcoming of look-up tables. LUTs generate output as soon as they're provided input. If all you had to work with was LUTs, your FPGA could perform all the Boolean algebra you might want, but your outputs would be determined solely based on the current inputs. The FPGA would know nothing about its past state. This would be very limiting. Implementing a counter would be impractical, since a counter requires knowledge of a previous value that can be incremented; so would storing the result of some math operation as a variable. Even something as critical as having a concept of time is impractical with just LUTs; you can only calculate values based on the now, not on anything in the past. The flip-flop enables these interesting capabilities, which is why it's critical to the operation of an FPGA.

How a Flip-Flop Works

A flip-flop stores state in the form of a high or low voltage, corresponding to a binary 1 or 0 or a true/false value. It does this by periodically checking the value on its input, passing that value along to its output, and holding it there. Consider the basic diagram of a *D flip-flop* shown in Figure 4-1. D flip-flops are the most common type of flip-flop in FPGAs, and they're the focus of this chapter. (I'll drop the *D* in front of *flip-flop* going forward.)

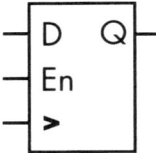

Figure 4-1: A diagram
of a D flip-flop

Notice that the component has three inputs on the left and one output on the right. The top-left input, labeled *D*, is the *data input* to the flip-flop. It's where data, in the form of 1s or 0s, comes in. The bottom-left input, labeled with what looks like a greater-than (>) sign, is the *clock input*, which synchronizes the performance of the flip-flop. At regular intervals, the clock input triggers the flip-flop to take the value from the data input and pass it to the output (labeled *Q* in the diagram).

The middle-left input, labeled *En*, is the *clock enable*. As long as the clock enable is high, the clock input will continue to trigger the flip-flop to update its output. If the clock enable input goes low, however, the flip-flop will ignore its clock and data inputs, essentially freezing its current output value.

To better understand how a flip-flop operates, we need to look more closely at the signal coming in to the clock input.

FLIP-FLOP COMPONENT TERMINOLOGY

The component presented in Figure 4-1, a flip-flop with a clock enable pin, isn't always called a flip-flop. FPGA manufacturers such as AMD and Intel do in fact use that terminology in their reference information, but a more technically accurate name is a *clocked D latch*. It's not valuable getting into the details about why one name is better than another; instead, for the purposes of this book, we'll use the real-world terminology that the FPGA manufacturers use and refer to these components as flip-flops.

The Clock Signal

A *clock signal*, often just called a *clock*, is a digital signal that steadily alternates between high and low, as shown in Figure 4-2. This signal is usually provided via a dedicated electronic component external to the FPGA. A clock is key to how FPGAs operate: it triggers other components, such as flip-flops, to perform their tasks. If you think of an FPGA as a set of gears, the clock is like the big gear that turns all the other gears. If the main gear isn't spinning, the others won't spin either. You could also think of the clock as the heart of the system, since it keeps the beat for the entire FPGA. Every flip-flop in the FPGA will be updated on the pulse of the clock's heartbeat.

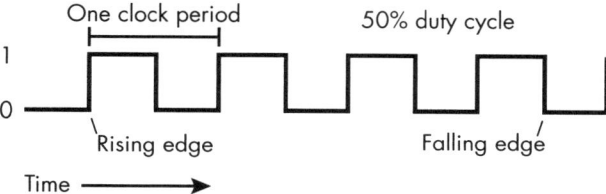

Figure 4-2: A clock signal

Notice the vertical lines in the clock signal diagram, where the signal jumps from low to high or high to low. These abrupt changes in the signal are called *edges*. When the clock goes from low to high, it's called a *rising edge*, and when it goes from high to low, it's called a *falling edge*. Flip-flops are conventionally triggered on each rising edge of the clock: whenever the clock signal changes from low to high, the flip-flop updates its output to match its data input.

NOTE   *It's possible to trigger a flip-flop with the falling edges of a clock, but this is* much *less common than using the rising edge.*

Every clock has a *duty cycle*, the fraction of time that the signal is high. For example, a signal with a 25 percent duty cycle is high one-quarter of the time and low three-quarters of the time. Almost all clocks, including the one shown in Figure 4-2, have a 50 percent duty cycle: they're half-on, half-off.

A clock also has a *frequency*, which is the number of repetitions from low to high and back again (called a cycle) in a second. Frequency is measured in hertz (Hz), or cycles per second. You may be familiar with your computer's CPU frequency, which can be measured in gigahertz (GHz), where 1 GHz is 1 billion Hz. FPGAs don't often run quite that quickly. More commonly, FPGA clock signals run in the tens to hundreds of megahertz (MHz), where 1 MHz is 1 million Hz. As an example, the clock on the Go Board (discussed in Appendix A) runs at 25 MHz, or 25 million cycles per second.

Another way to describe a clock's speed is to refer to its *period*, the duration of a single clock cycle. You can calculate the period by finding

$1 \div$ *frequency.* In the case of the Go Board, for instance, the clock period is 40 nanoseconds (ns).

A Flip-Flop in Action

A flip-flop operates on the transitions of its clock input. As mentioned previously, when a flip-flop sees a rising edge of the clock, it checks the state of the data input signal and replicates it at the output—assuming the clock enable pin is set to high. This process is called *registering,* as in, "the flip-flop *registers* the input data." Thanks to this terminology, a group of flip-flops is known as a *register,* and by extension, a single flip-flop can also be called a *one-bit register.* One flip-flop by itself is able to register a single bit of data.

To see how registering works in practice, we'll examine a few example inputs to a flip-flop and their corresponding outputs. First, consider Figure 4-3.

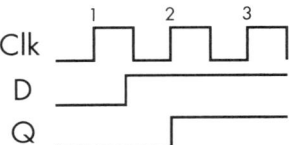

Figure 4-3: An example of flip-flop behavior

This figure shows three waveforms: the top one (Clk) represents an FPGA's clock signal, the middle one (D) is the data input of a flip-flop, and the bottom one (Q) is the flip-flop's output. Let's assume the clock enable is high, so the flip-flop is always enabled. We can see the waveforms across three cycles of the clock; the rising edge of each clock cycle is indicated with the numbers 1, 2, and 3. In between the first and second rising edges of the clock, the D input goes from low to high, but notice that the output doesn't immediately go high when the input does. Instead, it takes a bit of time for the flip-flop to register the change in the input. Specifically, it takes until the *next rising clock edge* for the flip-flop output to follow the input.

The flip-flop looks at the input data and makes the output match the input only at the rising edge of the clock, never between edges. In this case, at the rising edge of the second clock cycle, the output Q sees that D has gone from low to high. At this point, Q takes on the same value as D. On the third rising edge, Q again checks the value of D and registers it. Since D hasn't changed, Q stays high. Q also registered D at the rising edge of the first clock cycle, but since both D and Q were low at that point, Q didn't change.

Now consider Figure 4-4, which shows how a flip-flop responds to another example scenario.

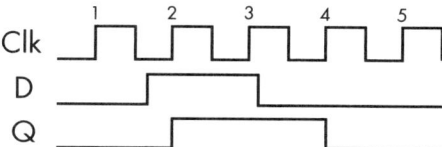

Figure 4-4: Another example of flip-flop behavior

Here we again see a flip-flop operating over several cycles of the clock. Again, let's assume the flip-flop is always enabled. Between the clock's first and second rising edges, input D goes from low to high. On the second rising edge, Q sees that D has gone high, so it toggles from low to high as well. On the third rising edge, Q sees D has stayed high, so it stays high, too. Between the third and fourth rising edges, D goes low, and the output similarly goes low on the fourth rising edge. On the last rising edge, D is still low, so Q stays low as well.

The previous examples have all assumed the clock enable input is high. Let's now show what happens when the flip-flop's clock enable isn't always high. Figure 4-5 shows the exact same Clk and D waveforms as Figure 4-4, but instead of the clock enable remaining high the whole time, it's only high at the third rising edge.

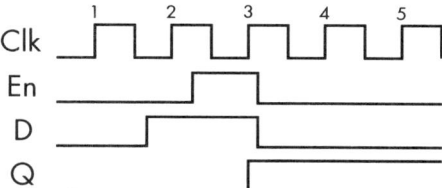

Figure 4-5: Flip-flop behavior with the clock enable signal

With the clock enable (En) now in play, a completely different output Q is generated. Q no longer "sees" that D has gone high on clock cycle two, since the clock enable is low at that point. Instead, Q only changes its output from low to high on clock cycle three, when the clock enable is high. On clock cycle four, D has gone low, but Q doesn't follow D. Instead, it stays high. This is because the clock enable has gone low at that point, locking the output in place. The flip-flop will no longer register any changes on D to Q.

These examples demonstrate flip-flop behavior, showing how a flip-flop's activity is coordinated by a clock. Additionally, we've seen how turning off the clock enable pin allows flip-flops to retain state, even when the input D is changing. This gives flip-flops the ability to store data for a long time.

A Chain of Flip-Flops

Flip-flops are commonly chained together, with the output from one flip-flop going directly into the data input of another flip-flop. For example, Figure 4-6 shows a chain of four flip-flops. For simplicity, let's assume these are always enabled.

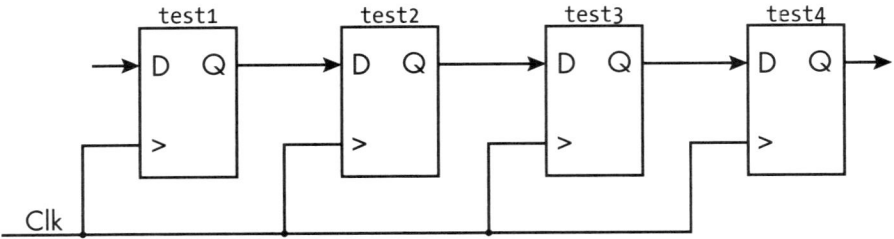

Figure 4-6: A chain of four flip-flops

The four flip-flops, labeled test1 through test4, are chained such that the output of test1 goes to the input of test2, the output of test2 goes to the input of test3, and so on. All four flip-flops are driven by the same clock. The clock synchronizes their operation: with each rising edge of the clock, all four flip-flops will check the value on their input and register that value to their output.

Suppose the test1 flip-flop registers a change at its input. Figure 4-7 illustrates how that change will propagate through the flip-flop chain, all the way to the output of test4.

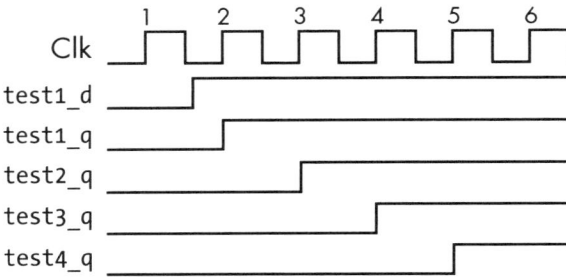

Figure 4-7: A change of input propagating through the flip-flop chain

The figure shows waveforms for the clock signal, the input and output of the test1 flip-flop (test1_d and test1_q, respectively), and the output of each subsequent flip-flop. On the first clock cycle rising edge (labeled 1), test1_d is low, so test1_q stays low as well. It's not until the second rising clock edge that the first flip-flop "sees" that the input has changed to high and registers that to its output. The test1 flip-flop's output is also the input to the test2 flip-flop, but notice that the output of test2 doesn't immediately change to high when the output of test1 does. Instead, test2_q changes one clock cycle later, on the third rising clock edge. Then, on the fourth rising edge, we see test3_q go high, and finally on the fifth rising edge test4_q goes high and stays high.

By adding three flip-flops behind test1, we've delayed the output by three clock cycles as the signal propagates through the chain. Each flip-flop in the chain adds a single clock cycle of delay. This technique of delaying signals by adding a chain of flip-flops is a useful design practice when working with FPGAs. Among other things, designers may chain flip-flops to create circuits that can delay or remember data for some amount of time, or to convert serial data to parallel data (or vice versa).

OTHER KINDS OF FLIP-FLOPS

In this chapter we're focusing on the D flip-flop. If you've taken a digital electronics course in college, there's a good chance your professor spent time talking about other types of flip-flops as well, including the *T flip-flop* and the *JK flip-flop*. In practice, however, you're unlikely to need to know anything about these other types of flip-flops to use an FPGA, as most FPGAs are made with D flip-flops. For this reason, I won't burden you with information about how the other kinds of flip-flops work, although it's important to acknowledge that they exist.

Project #3: Blinking an LED

Now that you know how flip-flops work, we'll make use of a couple of them in a project where the FPGA must remember information about its own state. Specifically, we're going to toggle the state of an LED each time a switch is released. If the LED was off before the switch is released, it should turn on, and if the LED was on, it should turn off.

This project uses two flip-flops. The first is for remembering the state of the LED: whether it's on or off. Without this memory, the FPGA would have no way of knowing whether to toggle the LED each time the switch is released; it won't know if the LED is on and needs to be turned off, or off and needs to be turned on.

The second flip-flop allows the FPGA to detect when the switch is released. Specifically, we're looking for the falling edge of the switch's electrical signal: its transition from high to low. A good way to look for a falling edge in an FPGA is to register the signal in question by passing it through a flip-flop. When the input value of the flip-flop (that is, the unregistered value) is equal to 0 but the previous output value (the registered value) is equal to 1, then we know that a falling edge has occurred. The falling edge of the switch is not to be confused with the rising edge of the clock; we're still using the rising edge of the clock to drive all of our flip-flops. Figure 4-8 shows the pattern to look for.

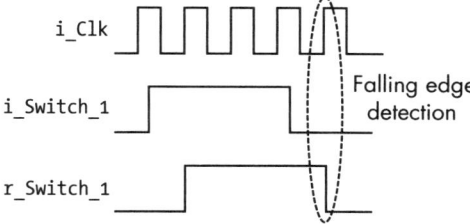

Figure 4-8: Falling edge detection using flip-flop

Here, i_Clk is the clock signal; i_Switch_1 represents the electrical signal from the switch, which passes into a flip-flop; and r_Switch_1 is the flip-flop's output. At the circled rising clock edge, we can see that i_Switch_1 is low, but r_Switch_1 is high. This pattern is how we can detect the falling edge of a signal. One thing to note is that while r_Switch_1 does go low on the rising clock edge, when the logic evaluates the state of r_Switch_1 at that same rising clock edge, it will still "see" that r_Switch_1 is high. Only after some small delay will the output of r_Switch_1 go low, following the state of i_Switch_1.

This project will also require some logic between the two flip-flops, which will be implemented in the form of a LUT. This will be your first glimpse of how flip-flops and LUTs work together in an FPGA to accomplish tasks. Figure 4-9 shows an overall block diagram for this project.

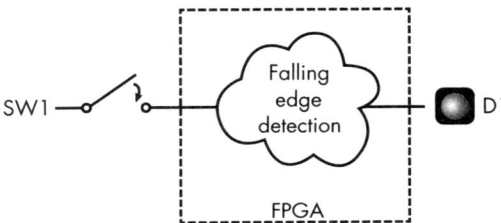

Figure 4-9: The Project #3 block diagram

The output of one of the switches on your development board (SW1) goes into the FPGA, where the falling edge detection logic is implemented. The output of this logic drives one of the board's LEDs (D1). Now we'll look at how to implement this design.

Writing the Code

We can write our LED-toggling code using Verilog or VHDL:

Verilog
```
module LED_Toggle_Project(
    input  i_Clk,
    input  i_Switch_1,
    output o_LED_1);

❶ reg r_LED_1    = 1'b0;
  reg r_Switch_1 = 1'b0;

❷ always @(posedge i_Clk)
  begin
  ❸ r_Switch_1 <= i_Switch_1;

  ❹ if (i_Switch_1 == 1'b0 && r_Switch_1 == 1'b1)
     begin
     ❺ r_LED_1 <= ~r_LED_1;
```

```
      end
    end

  assign o_LED_1 = r_LED_1;

endmodule
```

VHDL
```
library ieee;
use ieee.std_logic_1164.all;

entity LED_Toggle_Project is
  port (
    i_Clk      : in  std_logic;
    i_Switch_1 : in  std_logic;
    o_LED_1    : out std_logic
  );
end entity LED_Toggle_Project;

architecture RTL of LED_Toggle_Project is

❶ signal r_LED_1    : std_logic := '0';
  signal r_Switch_1 : std_logic := '0';

begin

❷ process (i_Clk) is
  begin
    if rising_edge(i_Clk) then
    ❸ r_Switch_1 <= i_Switch_1;
    ❹ if i_Switch_1 = '0' and r_Switch_1 = '1' then
      ❺ r_LED_1 <= not r_LED_1;
      end if;
    end if;
  end process;

  o_LED_1 <= r_LED_1;

end architecture RTL;
```

We begin by defining two inputs (the clock and the switch) and a single output (the LED). Then we create two signals ❶: r_LED_1 and r_Switch_1. We do this using the reg keyword (short for *register*) in Verilog, or the signal keyword in VHDL. Ultimately these signals will be implemented as flip-flops, or registers, so we prefix their names with the letter r. It's good practice to label any signals that you know will become registers r_*signal_name*, as it helps keep your code organized and easy to search.

Next, we initiate what's known as an always block in Verilog or a process block in VHDL ❷. This type of code block is triggered by changes in one or more signals, as specified by the code block's *sensitivity list*, which is given in parentheses when the block is declared. In this case, the block is sensitive to the clock signal, i_Clk. Specifically, this block will be triggered any time the clock changes from a 0 to a 1; that is, at each rising clock edge. Remember, when you use a clock to trigger logic within your FPGA, you'll almost always be using the clock's rising edges. In Verilog, we indicate this with the keyword posedge (short for *positive edge*, another term for *rising edge*) within the sensitivity list itself: always @(posedge i_Clk). In VHDL, however, we only put the signal name in the sensitivity list, and specify to watch for rising edges two lines later, with if rising_edge(i_Clk) then.

Within the always or process block, we create the first flip-flop of this project by taking the input signal i_Switch_1 and registering it into r_Switch_1 ❸. This line of code will generate a flip-flop with i_Switch_1 on the D input, r_Switch_1 on the Q output, and i_Clk going into the clock input. The output of this flip-flop will generate a one-clock-cycle delay of any changes to the input. This effectively gives us access to the previous state of the switch, which we need to know in order to detect the falling edge of the switch's signal.

We next check to see if the switch has been released ❹. To do this, we compare the current state of the switch with its previous state, using the flip-flop we just created ❸. If the current state (i_Switch_1) is 0 *and* the previous state (r_Switch_1) is 1, then we've detected a falling edge, meaning the switch has been released. The *and* check will be accomplished with a LUT.

At this point, perhaps you've noticed something surprising. First we assigned i_Switch_1 to r_Switch_1 ❸, then we checked if i_Switch_1 is 0 and r_Switch_1 is 1 ❹. You might think that since we just assigned i_Switch_1 to r_Switch_1, they'd always be equal, and the if statement would never be true. Right? Wrong! Assignments in an always or process block that use <= don't occur immediately. Instead, they take place on each rising edge of the clock and therefore *are all executed at the same time*. If at a rising clock edge i_Switch_1 is 0 and r_Switch_1 is 1, the if statement will evaluate as true, even as r_Switch_1 is simultaneously switching from a 1 to a 0 to match i_Switch_1.

Now we're thinking in parallel instead of serially! We've generated assignments that occur all at once, instead of one at a time. This is completely different from traditional programming languages like C and Python, where assignments occur one after the other. To further drive this point home, you could move the assignment of r_Switch_1 to the last line of the always or process block, and everything would still work the same. Formally, we call the <= assignment a *non-blocking assignment*, meaning it doesn't prevent ("block") other assignments from taking place at the same time. In Chapter 10, we'll revisit this concept and compare non-blocking assignments with blocking assignments.

Once we're inside the if statement, we toggle the state of the LED ❺. Doing so generates the second flip-flop used in this project. We take the current value of r_LED_1, invert it, and store the result back into the flip-flop. That might sound impossible, but it's perfectly valid. The output of the flip-flop will pass through a LUT, acting here as a NOT gate, and then be fed back into the flip-flop's input. This way, if the LED was on it'll turn off, and vice versa.

Adding Constraints

Once the code is ready, it's time to run the tools to build the FPGA image and program your board. First, since this project uses a clock, you need to add a constraint telling the FPGA tool about the clock's period. The clock period tells the timing tool how much time is available to route wires between flip-flops. As clock speed increases, it gets harder for the FPGA to *meet timing*, or achieve all the desired tasks within each clock cycle. For slower clocks, with frequencies on the order of tens of megahertz, you shouldn't have any problems meeting timing. In general, it's only when you deal with clocks that are faster than 100 MHz that you may start to run into timing issues.

The clock period will vary from one development board to another, and can be found in your board's documentation. To tell Lattice iCEcube2 about the clock period, create a new text file with a *.sdc* file extension containing something like the following:

```
create_clock  -period 40.00 -name {i_Clk} [get_ports {i_Clk}]
```

This creates a clock with a 40 ns period (25 MHz frequency) and assigns that constraint to the signal called i_Clk in your design. This constraint will work for the Go Board, as an example, but if your board has a different clock period, replace 40.00 with the appropriate value.

Right-click **Constraint Files** under **Synthesis Tool** and select the *.sdc* file to add it to your project in iCEcube2. Remember from Chapter 2 that we previously had a single *.pcf* constraint file telling the tools which signals to map to which pins. Now we have an additional constraint file just for the clock. Both are critical for getting your FPGA to work correctly.

We also need to update the *.pcf* file to include the pin corresponding to the new clock signal. On the Go Board, for example, the clock is connected to pin 15 of the FPGA, so you would need to add the following pin constraint:

```
set_io i_Clk 15
```

Check the schematic for your development board to see which pin has the clock as an input, and replace the 15 as appropriate.

Building and Programming the FPGA

You're now ready to run the build. When you do this, the tools will generate some reports. The synthesis report should look something like this:

```
--snip--
Resource Usage Report for LED_Toggle_Project

Mapping to part: ice40hx1kvq100
Cell usage:
SB_DFF 2 uses
SB_LUT4 1 use

I/O ports: 3
I/O primitives: 3
```

```
   SB_GB_IO 1 use
   SB_IO 2 uses

   I/O Register bits: 0
❶ Register bits not including I/Os: 2 (0%)
   Total load per clock:
    ❷ LED_Toggle_Project|i_Clk: 1
   Mapping Summary:
❸ Total LUTs: 1 (0%)
```

This report tells us that we're using two register bits ❶, meaning our design includes two flip-flops. This is exactly what we expected. The report also shows that we're using one LUT ❸. This single LUT will be able to perform both the AND and NOT operations required in the code. Notice, too, that the tools identified the signal i_Clk as a clock ❷.

Now let's look at the place and route reports, which you can view in iCEcube2 by going to **P&R Flow ▸ Output Files ▸ Reports**. There are two reports here. The first is a pin report, which tells you which signals were mapped to which pins. You can use this to confirm that your signals were mapped correctly. The second is the timing report. It has a section labeled "Clock Frequency Summary" that should look something like this:

```
--snip--
 1::Clock Frequency Summary
========================================================
Number of clocks: 1
Clock: i_Clk | Frequency: 654.05 MHz | Target: 25.00 MHz |
--snip--
```

This section tells you if the constraint file was accepted correctly. Here we see that the tools have found our clock, i_Clk. The Target property indicates the tools have recognized a 25 MHz constraint placed on the clock (your number will vary, depending on your development board), while the Frequency property tells us the maximum frequency at which the FPGA could theoretically run our code successfully. In this case, we could run this FPGA at 654.05 MHz and it would still be guaranteed to work correctly. That's quite fast! As long as the Frequency property is higher than the Target property, you shouldn't have any issues running your code. A problem would show up here in the form of a *timing error*, which happens when the target clock speed is greater than the frequency that the tools can achieve. In Chapter 7, we'll take a deeper look at what causes timing errors and how to fix them.

Now that you've successfully built the FPGA design, you can program your board and test the project. Try pushing the switch several times. You should see the LED toggle on or off each time the switch is released. Congratulations, you've got your first flip-flop working!

However, you may notice something strange going on. The LED may not appear to change its state with each release. You might think that the FPGA isn't registering the releases of the switch, but in fact the LED is toggling two or more times with each release, so quickly that your eyes don't see it. The cause is related to the physical workings of the switch itself. To solve this issue, the switch needs to be *debounced*. You'll learn what this means and how to do it in the next chapter.

Combinational Logic vs. Sequential Logic

There are two kinds of logic that can take place inside an FPGA: combinational logic and sequential logic. *Combinational logic* is logic for which the outputs are determined by the present inputs, with no memory of the previous state. This kind of logic is achieved with LUTs, which you'll recall generate their output based only on their current inputs. *Sequential logic*, on the other hand, is logic for which the outputs are determined both by present inputs and previous outputs. Sequential logic is achieved with flip-flops, since flip-flops don't immediately register changes on their inputs to their outputs, but rather wait until the rising edge of the clock to act on the new input data.

NOTE *You may also see combinational logic and sequential logic referred to as* combinatorial logic *and* synchronous logic, *respectively.*

It might not be obvious that a flip-flop's output depends on its previous output, so let's explore an example to make this more concrete. Suppose the flip-flop is enabled, its input is low, its clock is low, and the output is low. Then suddenly the input goes high, then back low again quickly. What will the output do? Nothing! It stays low, since there was no clock edge to trigger a change. Now, what happens if that same flip-flop has the same initial conditions, except the output is high? In this case, of course, the output will stay high. But if we only looked at the inputs (D, En, and Clk), *we would be unable to predict the output state.* You need to know what the output of the flip-flop was (its previous state) to determine the flip-flop's current state. That's why a flip-flop is sequential.

Knowing if your code is going to instantiate LUTs (combinational logic) or flip-flops (sequential logic) is critical to being a good FPGA designer, but sometimes it can be hard to tell the difference. In particular, an always block (in Verilog) or process block (in VHDL) can define a block of either combinational logic or sequential logic. We'll consider examples of each to see how they differ.

First, here's an example of a combinational implementation in Verilog and VHDL:

Verilog
```
always @ (input_1 or input_2)
  begin
    and_gate <= input_1 & input_2;
  end
```

```
process (input_1, input_2)
begin
   and_gate <= input_1 and input_2;
end process;
```

Here we've created an `always` or `process` block with a sensitivity list (the signals in the parentheses) that includes two signals: `input_1` and `input_2`. The code block performs an AND operation on the two signals.

This block of Verilog or VHDL code will only generate LUTs; it won't generate any flip-flops. For our purposes, flip-flops require a clock input, and there is no clock. Since no flip-flops are generated, this is combinational logic.

Now consider a slight modification to the examples just shown:

Verilog
```
always @ (posedge i_Clk)
   begin
      and_gate <= input_1 & input_2;
   end
```

VHDL
```
process (i_Clk)
begin
   if rising_edge(i_Clk) then
      and_gate <= input_1 and input_2;
   end if;
end process;
```

This code looks very similar to the previous examples, except now the `always` or `process` block's sensitivity list has changed to be sensitive to the signal `i_Clk`. Since the block is sensitive to a clock, it's now considered sequential logic. This block will actually still require a LUT to perform the AND operation, but in addition to that the output will utilize a flip-flop, since the clock is gating the output from updating all the time.

While all the examples in this section are valid code, I'm going to make a suggestion, especially for FPGA beginners: when writing your code, only create *sequential* `always` blocks (in Verilog) or `process` blocks (in VHDL). The way to do this is to ensure that the block's sensitivity list only has a clock in it. (A clock and a reset is OK too, as we'll discuss later in the chapter.) Combinational `always` blocks and `process` blocks can get you into trouble: you can generate a latch by accident. We'll explore latches in the next section, but basically, they're bad. Additionally, I find code is more readable if you know that every time you come across an `always` block or `process` block, it will always be generating sequential logic.

As for combinational-only logic, write it outside of an `always` block or `process` block. In Verilog, the keyword `assign` is useful. In VHDL, you can simply use the <= assignment to create combinational logic.

The Dangers of Latches

A *latch* is a digital component that can store state without the use of a clock. In this way, latches perform a similar function as flip-flops (namely, storing state), but the method they use is different since there's no clock involved. Latches are dangerous and can be inadvertently generated when working with combinational code. In my career, I've never once generated a latch *on purpose*, only by accident. It's highly unlikely that you'd ever actually want to generate a latch either, so it's important to understand how to avoid them.

You always want your FPGA designs to be predictable. Latches are dangerous because they violate this principle. FPGA tools have a very difficult time understanding the timing relationship of a latch and how other components connected to it will perform. If you do manage to create a latch with your code, the FPGA tools will scream at you with warnings about the fact that you've done a horrible thing. Please don't ignore these warnings.

So how can this happen? A latch is created when you write a combinational process block or conditional assignment (in VHDL) or a combinational always block (in Verilog) with an *incomplete assignment*, meaning the output isn't assigned under all possible input conditions. This is bad and should be avoided. Table 4-1 shows an example of a truth table that would generate a latch.

Table 4-1: A Truth Table That Creates a Latch

Input A	Input B	Output Q
0	0	0
0	1	1
1	0	1
1	1	Undefined

This truth table has two inputs and one output. The output is 0 when both inputs are 0, and it's 1 when input A is 0 and input B is 1, or when input A is 1 and input B is 0. But what happens when both inputs are 1? We haven't explicitly stated what will occur. In this case, the FPGA tools assume that the output should retain its previous state, much like a flip-flop is capable of doing, but without the use of a clock. For example, if the output is 0 and both inputs go high, the output will stay 0. If the output is 1 and both inputs go high, the output will stay 1. This is the behavior that a latch creates: the ability to store state without a clock.

Let's take a look at how this truth table could be created in Verilog and VHDL. Don't write code like this!

Verilog ❶
```
always @ (i_A or i_B)
begin
  if (i_A == 1'b0 && i_B == 1'b0)
    o_Q <= 1'b0;
```

```
        else if (i_A == 1'b0 && i_B == 1'b1)
          o_Q <= 1'b1;
        else if (i_A == 1'b1 && i_B == 1'b0)
          o_Q <= 1'b1;
    ❷ // Missing one last ELSE statement!
    end
```

VHDL ❶
```
process (i_A, i_B)
begin
  if i_A = '0' and i_B = '0' then
    o_Q <= '0';
  elsif i_A = '0' and i_B = '1' then
    o_Q <= '1';
  elsif i_A = '1' and i_B = '0' then
    o_Q <= '1';
❷ -- Missing one last ELSE statement!
  end if;
end process;
```

Here, our always or process block is combinational because there's no clock in the sensitivity list ❶ or the block itself, just two inputs, i_A and i_B. We mimic the incomplete truth table assignment of the output o_Q using conditional checks. Notice that we don't explicitly check the condition where i_A and i_B are both 1. Big mistake!

If you were to try to synthesize this faulty code, the FPGA tools would generate a latch and warn you about it in the synthesis report. The warning would look something like this:

```
@W: CL118 :"C:\Test.v":8:4:8:5|Latch generated from always block for signal
o_Q; possible missing assignment in an if or case statement.
```

The tools are pretty good. They tell you that there's a latch, they tell you which signal it is (o_Q), and they tell you why it might be occurring.

To avoid generating a latch, we could add an else statement ❷, which will cover all remaining possibilities. As long as the output is defined for all possible inputs, we'll be safe. An even better solution, however, would be not to use a combinational always or process block at all. I discourage the use of combinational always or process blocks precisely because it's easy to make this mistake of omitting an else statement. Instead, we can use a sequential always or process block. Here's what that looks like:

Verilog ❶
```
always @ (posedge i_Clk)
begin
  if (i_A == 1'b0 && i_B == 1'b0)
    o_Q <= 1'b0;
  else if (i_A == 1'b0 && i_B == 1'b1)
    o_Q <= 1'b1;
  else if (i_A == 1'b1 && i_B == 1'b0)
    o_Q <= 1'b1;
end
```

VHDL ❶
```
process (i_Clk)
begin
  if rising_edge(i_Clk) then
    if i_A = '0' and i_B = '0' then
      o_Q <= '0';
    elsif i_A = '0' and i_B = '1' then
      o_Q <= '1';
    elsif i_A = '1' and i_B = '0' then
      o_Q <= '1';
    end if;
  end if;
end process;
```

We now have a sequential always or process block, because we're using a clock in the sensitivity list ❶ and within the block itself. As a result, o_Q will create a flip-flop rather than a latch. Flip-flops don't have the same unpredictable timing issues that latches do. Remember that the flip-flop can utilize its en input to retain a value. The flip-flop's en input will be disabled when i_A and i_B are both high. This will retain the flip-flop's output with whatever state it had previously, performing the same behavior as the latch, but in a safe, predictable way.

One side effect of switching to a sequential always or process block is that it now takes a single clock cycle for the output to be updated. If it's critical that this logic be combinational—with the output updating as soon as one of the inputs changes, with no clock delay—then you need to ensure that the output is specified for all possible input conditions.

There's one other way to generate latches in VHDL. VHDL has the keyword when, which can be used in a conditional assignment. Verilog has no equivalent syntax, so this code snippet is for VHDL only:

```
o_Q <= '0' when (i_A = '0' and i_B = '0') else
       '1' when (i_A = '0' and i_B = '1') else
       '1' when (i_A = '1' and i_B = '0');
```

This code exists outside of a process block, and again we haven't explicitly stated what o_Q should be assigned to when i_A and i_B are both 1, so the FPGA tools will infer a latch here. The latch will enable the output to keep its previous state, but that's likely not what we intended. Instead, we should be specific with our code and ensure that we have an else condition that sets o_Q for all possible inputs.

Resetting a Flip-Flop

Flip-flops have an additional input that we haven't discussed yet, called *set/reset*, or often just *reset*. This pin resets the flip-flop back to an initial state, which could be 0 or 1. Resetting flip-flops is useful when the FPGA first powers up and initializes. For example, you might want to reset your flip-flops that control a state machine to the initial state (we'll discuss state machines in Chapter 8). You might also want to reset a counter to some initial value, or reset a filter back to zero. Resetting flip-flops is one method to ensure your flip-flops are in a specific state prior to operation.

There are two types of resets: synchronous and asynchronous. *Synchronous resets* occur at the same time as the clock edge, whereas *asynchronous resets* can occur at any time. You might trigger an asynchronous reset with a button press external to the FPGA, for example, since the button press can come at any point in time. Let's look at how to code a reset, starting with a synchronous one:

Verilog
```
❶ always @ (posedge i_Clk)
   begin
❷ if (i_Reset)
       o_Q <= 1'b1;
❸ else
   --snip--
```

VHDL
```
❶ process (i_Clk)
   begin
     if rising_edge(i_Clk) then
❷ if i_Reset = '1' then
         o_Q <= '1';
❸ else
   --snip--
```

Here we have an `always` or `process` block with a normal sensitivity list; it's only sensitive to changes of the clock ❶. Inside the block, we first check the state of i_Reset ❷. If it's high, then we reset the signal o_Q to 1. This is our synchronous reset, since it's happening on the edge of the clock. If i_Reset is low, we proceed with the else branch of the block ❸, where we'd write whatever code we want to be executed under normal operating (non-reset) conditions.

Notice that in this example we're checking if the reset is high. Sometimes resets can be active low, however, which is usually indicated by _L or _n at the end of the signal name. If this were an active low reset, we would check for the signal being 0 rather than 1.

Now let's take a look at an asynchronous reset:

Verilog
```
❶ always @ (posedge i_Clk or i_Reset)
   begin
❷ if (i_Reset)
       o_Q <= 1'b1;
❸ else
   --snip--
```

VHDL
```
❶ process (i_Clk, i_Reset)
   begin
❷ if (i_Reset = '1') then
       o_Q <= '1';
```

```
❸ elsif rising_edge(i_Clk) then
--snip--
```

Notice that we've added i_Reset into the always or process block's sensitivity list ❶. Now, rather than checking the clock state first, we check the reset state first ❷. If it's high, then we perform whatever reset conditions we want, in this case setting o_Q to 1. Otherwise, we proceed normally ❸.

The choice between synchronous and asynchronous resets should be documented in the user guide for your specific FPGA—some FPGAs are optimized to handle one or the other. Additionally, resets can create strange bugs if they're not treated properly. Therefore, I strongly recommend consulting the documentation to make sure you're resetting flip-flops correctly for your device.

Look-Up Tables and Flip-Flops on a Real FPGA

Now you understand that LUTs and flip-flops exist on FPGAs, but they may still seem a bit abstract. To get a more concrete picture, let's look at how LUTs and flip-flops are actually wired together in a real FPGA. The image in Figure 4-10 is taken from the datasheet for the Lattice iCE40 LP/HX family of FPGAs, the type of FPGA compatible with iCEcube2.

Datasheets are used throughout the electronics industry to explain the details of how a component works. Each FPGA will have at least a few unique datasheets with different pieces of information, and more complicated FPGAs can have dozens of them.

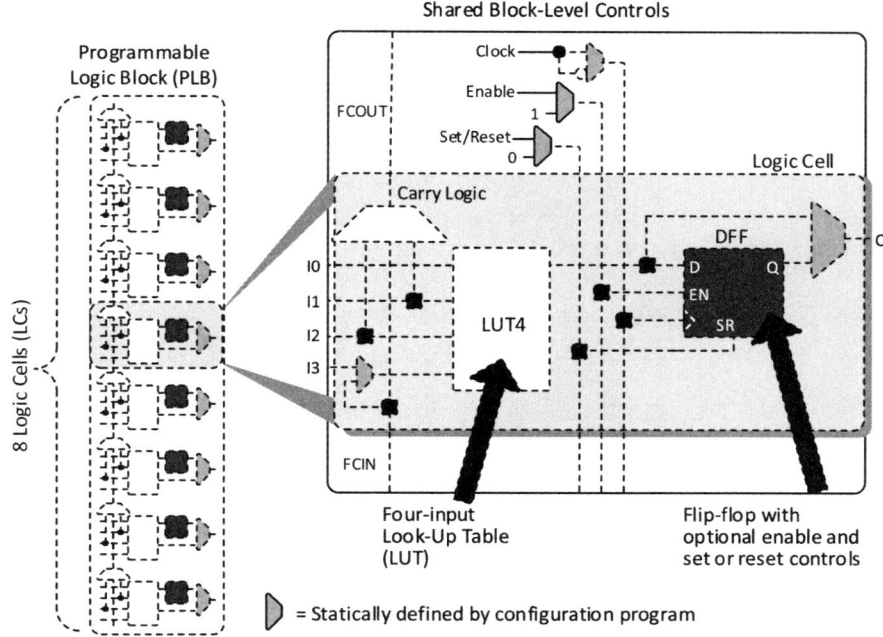

Figure 4-10: LUTs and flip-flops in a real FPGA

Every FPGA, whether from Lattice, AMD, Intel, or whoever else, will have an image very similar to Figure 4-10 in its specific family datasheet. This particular image shows the basic building block of Lattice iCE40 FPGAs, which Lattice calls the *Programmable Logic Block (PLB)*. Each FPGA company has its own unique name for these basic building blocks; for example, AMD calls them *Configurable Logic Blocks (CLBs)*, while Intel uses *Adaptive Logic Modules (ALMs)*. We'll look at the details of how the PLB from Lattice works as an example.

Looking at the left side of the image, we see there are eight logic cells in each PLB. The right side shows a zoomed-in version of a single logic cell. Inside it, notice that there's a rectangle labeled LUT4. This is a four-input look-up table! There's also a dark gray box labeled DFF. This is a D flip-flop! The LUT and the flip-flop truly are the two most critical components inside an FPGA.

This diagram is telling us that at the most fundamental level there's one LUT and one flip-flop inside each logic cell, and there are eight logic cells in a PLB. The PLB is copy-pasted hundreds or thousands of times inside the FPGA to provide enough LUTs and flip-flops to do all the required work.

On the left side of the DFF component (the flip-flop), notice the same three inputs we originally saw in Figure 4-1: data (D), clock enable (EN), and clock (>). The fourth input at the bottom of the component is the set/reset (SR) input we discussed in the previous section.

As you've seen, the clock enable input allows the flip-flop to keep its output state for multiple clock cycles. Without the En input, the output would just follow the input with one clock cycle of delay. Adding the En input lets the flip-flop store a state for a longer duration.

The last thing to notice in the diagram is the carry logic block, shown above and to the left of the LUT4. This block is mostly used to speed up arithmetic functions, such as addition, subtraction, and comparison.

While reviewing this diagram gave us an interesting look inside an FPGA and highlighted the central role of the LUT and the flip-flop, it isn't critical to memorize every detail of the PLB's architecture. You don't need to remember all the connections and how each is wired to its neighbor. In the real world, you write your Verilog or VHDL, and the FPGA tools take care of mapping that code onto the FPGA's resources. This is particularly useful if you want to switch from one type of FPGA to another (say, from a Lattice to an AMD). The beauty of Verilog and VHDL is that the code is generally portable; the same code works on different FPGAs, provided they have enough LUTs and flip-flops to do what you want.

Summary

In this chapter you learned about the flip-flop, which, along with the LUT, is one of the two most important components in an FPGA. You saw how flip-flops allow FPGAs to keep state, or remember past values, by only registering data from the input to the output on the positive edges of a clock signal. You learned how logic driven by flip-flops and clock signals is sequential, in contrast to the combinational logic of LUTs, and you got your first glimpse of how flip-flops and LUTs work together through a project toggling an LED. You also learned how to avoid generating latches and how to reset a flip-flop to a default state.

In future chapters, as you build more complex blocks of code, you'll become more familiar with how flip-flops and LUTs interact and see how you can use just these two kinds of components to create large, sophisticated FPGA designs. You'll also see the role flip-flops play in keeping track of counters and state machines.

5

TESTING YOUR CODE WITH SIMULATION

There are two ways to find the bugs that will inevitably arise in an FPGA design. The first is to program the FPGA, run it, and see what happens. This is called finding bugs *on hardware.* The other way is to use a computer to inject test cases into your FPGA code to see how the code responds *before* you actually program the FPGA. This is called finding bugs *in simulation.*

For very simple projects, such as the ones we've explored so far in this book, jumping straight to programming the FPGA without any kind of simulation may be a reasonable approach (and it's the one we've taken up to this point). However, as your FPGA designs grow more complicated, finding bugs on hardware becomes incredibly difficult. In nearly all cases, it's significantly easier to find bugs in simulation. After thoroughly simulating and debugging a design, there's nothing more satisfying than finally programming your FPGA and having everything work perfectly the first time.

In this chapter, you'll learn how simulation works and see why it's an essential step in the FPGA design process. We'll explore a free simulator tool and I'll introduce the testbench to show how you can write test code to stress your design. You'll try out these concepts by adding a debounce circuit to the LED-toggling project from the previous chapter (Project #3) and simulating the design. Finally, we'll take a look at verification, which is a more formal and rigorous process for testing out FPGA and ASIC designs.

Why Simulation Matters

Simulation is important because your FPGA is essentially a black box, as shown in Figure 5-1. When you program the FPGA, you're able to change the inputs and see how the outputs respond, but you're unable to see the details of what's going on inside the box itself. You can't follow the individual variables and data signals as they flow inside your FPGA.

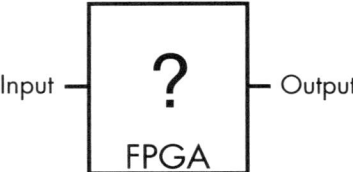

Figure 5-1: What's in the box?

If something goes wrong inside that black box (and it will), and the output isn't what you expect, figuring out the problem is very difficult. The solution is to use a computer to simulate the inner workings of the black box in a way you can follow. Simulation effectively opens up the black box of your FPGA so you can see what's going on inside.

Let me give you an example of how useful this can be. In a past job, I had a coworker who was trying to fix a problem with his FPGA design. For some reason the data was getting mixed up inside the FPGA. He spent weeks using oscilloscopes and logic analyzers to send data off the FPGA so he could try to find where the issue was coming from. At one point I asked him if he had simulated the design at all. He had not: he didn't have any experience with simulation and didn't feel he could take the time out to learn. I checked his code out of revision control and put together a simulation for it, and within a few hours had found the problem.

Simulating your design allows you to stress it to see how it reacts. In this case, I was able to re-create the exact failure in simulation and fix the issue very quickly. Ironically, in the time my coworker had spent attempting to debug the problem on hardware, he could easily have learned how to do the simulation himself. It's an even more attractive option when you consider that once you know how simulation works, you can use that knowledge again and again.

FPGA Simulation Tools

There are several popular FPGA simulation tools available. FPGA build tools often have a simulation tool bundled with them, in one large downloadable package; FPGA companies know that their designers want to run simulations, and they want to make it easy to do that. These tools are usually free and convenient, but they can be many gigabytes in size and their complexity can be overwhelming for beginners.

An alternative solution to the large FPGA tools is to use a standalone simulator. The benefit to this is that if you switch from Intel (Altera) to AMD (Xilinx), for example, you don't need to learn a whole new tool; your simulator can stay the same. There are two popular standalone simulation tools that I generally recommend: ModelSim and EDA Playground. ModelSim is probably the most popular commercial simulator. It can be downloaded and installed on Windows and Linux. A full license is expensive, costing around $2,000, but a free version with limited features is available.

EDA Playground, by contrast, is a freely available web-based simulator. I recommend using it when you're first learning about FPGA design for a few reasons. First, it's free. Second, since it's web-based, there's no download required. Finally, EDA Playground allows you to share your code with others via a web link. For the purposes of this book, we'll focus on this tool.

To get started with EDA Playground, first navigate to *https://edaplayground .com*. To run simulations and save your progress, you'll need to create an account and log in. Once you do so, you should see a screen like the one in Figure 5-2.

Figure 5-2: The EDA Playground main screen

Notice that there are two main code windows. The window on the right, titled *design.sv*, is where the FPGA design code you want to test goes. This code is typically called the *unit under test (UUT)* or *device under test (DUT)*. The window on the left, called *testbench.sv*, is where you write *testbenches*, code that will exercise your FPGA design during simulation. We'll discuss how testbenches work in the next section.

By default, EDA Playground is configured for SystemVerilog/Verilog designs, which is why the two window labels have *.sv* (SystemVerilog) file extensions. If you wish to reconfigure EDA Playground for VHDL, select **VHDL** in the drop-down menu under Testbench + Design on the left side of the window.

SYSTEMVERILOG

SystemVerilog is a superset of Verilog, meaning it has all the features of Verilog *and more*. Anything you write with Verilog will also work with SystemVerilog. SystemVerilog is often used in simulation, even of VHDL designs. There are a few reasons for this. For one, SystemVerilog offers many high-level language features, such as classes and interfaces, that make creating your testbench code easier and make that code more reusable across projects. In addition, it provides comprehensive support for assertions, which is a significant advantage when creating simulations. In VHDL and vanilla Verilog, assertions are much more limited.

Before you can run code in EDA Playground, you'll need to select a simulator tool. This is the actual product that will run your code. You can play around with different tools listed in the drop-down menu under Tools & Simulators to see if you prefer one over another. In general, I find that they behave similarly, though some are exclusively for Verilog or VHDL. I've had good luck using Mentor Questa or Aldec Riviera.

Another neat feature of EDA Playground is the Examples section of the toolbar. Here, you can explore sample testbenches that have been made freely available. You can see how they work and modify them for your own experiments, and perhaps gain some insights into clever ways to write your own code.

The Testbench

The purpose of a testbench is to exercise your UUT in a simulation environment so you can analyze it and see if it's behaving as expected. The testbench code instantiates the UUT. As you can see in Figure 5-3, the testbench provides all the required inputs to the UUT and monitors all the outputs.

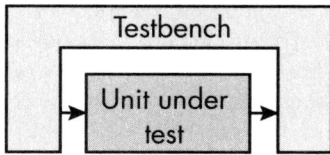

Figure 5-3: A testbench exercises
a UUT so you can analyze it.

If your UUT has a clock as an input, for example, the testbench will
need to generate that clock and feed it in to the UUT. Similarly, if there's
a data interface into your UUT, the testbench will likely need to generate
some sample data to supply to that interface. The testbench monitors all
the outputs from the UUT, allowing it to see how the UUT responds to the
input data. During the simulation, you'll also be able to dive into the UUT
itself to see how all of its internal signals are behaving in response to the
testbench's inputs. You can monitor every element of your design—every
register, clock, wire, memory, and so on—and make sure they're all behaving as intended.

Writing a Testbench

Let's take a look at a simple example by writing a testbench for the AND
gate project from Chapter 3 (Project #2). First, to review, here's the original
project code that we want to test:

Verilog

```
module And_Gate_Project
  (input i_Switch_1,
   input i_Switch_2,
   output o_LED_1);

assign o_LED_1 = i_Switch_1 & i_Switch_2;

endmodule
```

VHDL

```
library ieee;
use ieee.std_logic_1164.all;

entity And_Gate_Project is
  port (
    i_Switch_1 : in std_logic;
    i_Switch_2 : in std_logic;
    o_LED_1    : out std_logic);
end entity And_Gate_Project;

architecture RTL of And_Gate_Project is
begin
  o_LED_1 <= i_Switch_1 and i_Switch_2;
end RTL;
```

Enter the module or entity's code into the *design.sv* or *design.vhd* window on the right side of EDA Playground. To test the code completely, we want to exercise it to make sure that the output behaves as intended with all possible input combinations. In this case, the total range of input combinations is pretty small: since there are two inputs, there are just four possible combinations to test in order to fully exercise the UUT. We'll create a testbench in Verilog and VHDL to instantiate the UUT and test it by passing in each of the four input combinations. Enter the following code into the *testbench.sv* or *testbench.vhd* window in EDA Playground:

Verilog ❶
```
module And_Gate_TB();

   reg r_In1, r_In2;
   wire w_Out;

❷ And_Gate_Project UUT
   (.i_Switch_1(r_In1),
    .i_Switch_2(r_In2),
    .o_LED_1(w_Out));

❸ initial
     begin
     ❹ $dumpfile("dump.vcd"); $dumpvars;
       r_In1 <= 1'b0;
       r_In2 <= 1'b0;
       #10;
       r_In1 <= 1'b0;
       r_In2 <= 1'b1;
       #10;
       r_In1 <= 1'b1;
       r_In2 <= 1'b0;
       #10;
       r_In1 <= 1'b1;
       r_In2 <= 1'b1;
       #10;
       $finish();
     end
endmodule
```

VHDL
```
library IEEE;
use IEEE.std_logic_1164.all;
use std.env.finish;

❶ entity And_Gate_TB is
   end entity And_Gate_TB;

   architecture behave of And_Gate_TB is
     signal r_In1, r_In2, w_Out : std_logic;
   begin

   ❷ UUT : entity work.And_Gate_Project
       port map (
```

```
      i_Switch_1 => r_In1,
      i_Switch_2 => r_In2,
      o_LED_1    => w_Out);

❸ process is
  begin
    r_In1 <= '0';
    r_In2 <= '0';
    wait for 10 ns;
    r_In1 <= '0';
    r_In2 <= '1';
    wait for 10 ns;
    r_In1 <= '1';
    r_In2 <= '0';
    wait for 10 ns;
    r_In1 <= '1';
    r_In2 <= '1';
    wait for 10 ns;
    wait for 10 ns;
    finish;
  end process;
end behave;
```

First, notice that this is the first time we've seen a Verilog module or VHDL entity that has no inputs or outputs declared ❶. This is because this testbench doesn't connect to any external signals; as you saw earlier, in Figure 5-3, the testbench itself provides the inputs.

Inside the module/entity, we instantiate the UUT ❷. We connect the inputs of the UUT to r_In1 and r_In2, signals that we declare in the testbench. These signals will be the stimuli provided to see how the UUT responds. We'll be monitoring the output, w_Out, to see how it reacts to changing inputs. I like to use the w_ prefix on signal names to represent wires, or interconnections within the FPGA. Remember, we want to make sure the AND gate is working as expected.

We start driving the stimuli (inputs) within an initial block in Verilog or a process block in VHDL ❸. This block will start at the beginning of the simulation and will execute from top to bottom in sequence. We send each of the four possible input combinations to the UUT, one after the other. Using delay statements, we add a 10 ns pause between each input combination to allow time for the simulation to update the w_Out signal after each change. In Verilog we use the #10 delay feature, and in VHDL we use wait for 10 ns;. As you'll see later in this chapter, these time-based delays—indeed, any reference to the passage of time—are non-synthesizable, meaning they would not work on an actual FPGA; however, they work perfectly well in simulation.

In the Verilog version, note that EDA Playground requires the $dumpfile directive ❹. This allows the simulator to generate waveforms, which we'll cover in the next section. This line isn't required in VHDL.

Running a Testbench and Viewing Waveforms

Running a testbench generates *waveforms*, or visual representations of the signals in your test environment, showing you how they change over time. Waveforms are a powerful tool for investigating failures in an FPGA design during simulation; the more you work with FPGAs, the more time you'll spend staring at waveforms. EDA Playground makes examining waveforms easy with its built-in waveform viewer, EPWave.

Let's run our AND gate testbench and view the resulting waveform in EPWave. First, check the **Open EPWave After Run** checkbox in the Tools & Simulators section of the toolbar on the left side of the EDA Playground window. If you're using VHDL, you will need to specify which entity is the top of your design. To do that, enter **And_Gate_TB** in the Top Entity dialog. Then choose a simulator tool from the drop-down menu and hit **Run**. Figure 5-4 shows the resulting waveform.

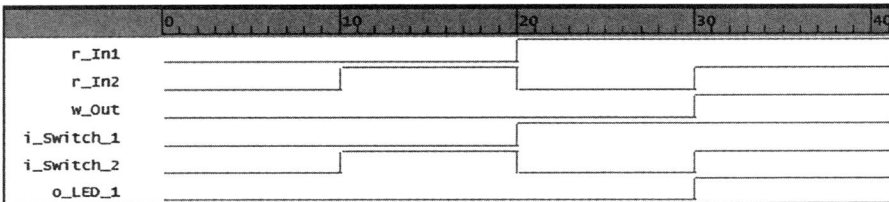

Figure 5-4: The AND gate testbench waveform output

Here we see all the signals that exist in the design, and we can note the time in nanoseconds when each signal changes from high to low or low to high. The top three signals (r_In1, r_In2, w_Out) are the testbench signals. The bottom three (i_Switch_1, i_Switch_2, o_LED_1) are in the UUT. Since we wired the testbench and UUT signals together when we instantiated the UUT, the corresponding testbench/UUT signals look the same. For example, r_In1 has the same waveform as i_Switch_1. Note that if the UUT had other internal signals that weren't brought out of the module, you would be able to see waveforms for those as well, and they wouldn't have a corresponding testbench signal.

Looking at the waveform, we can see that the UUT is working as expected. The AND gate output (o_LED_1 and w_Out) is high only when both inputs are also high. When only one input is high, or when both inputs are low, the output is low. As you examine the waveform, take a look back at the testbench code and notice how the changes in the waveform correspond to the statements in the initial or process block. In the code, for example, both inputs start out low, and then r_In2 goes high after a 10 ns pause. Looking at the 10 ns mark in the waveform, you can see that this is where r_In2 and i_Switch_2 change from low to high.

Although this was a simple example, it illustrates the power of the testbench to simulate your FPGA design and let you see everything that is happening. You can monitor all the interactions within the design, and if a signal isn't behaving as expected, you can investigate why that is, modify your code, and run the testbench again to generate a new waveform. Often when debugging issues, I'll rerun simulations dozens of times until my design is behaving as desired.

In this case, since we were testing a single basic module, we were able to evaluate everything using a single testbench file. For more complicated simulations, however, testbenches can contain many different files that all work together to simulate, monitor, and check your design to make sure it's behaving as intended.

You'll see how testbenches work in more detail in our next project, where we'll write a testbench and simulate the FPGA design prior to programming the hardware. This will help you gain confidence that your code is working, and it will allow you to identify and fix any bugs early in the process. The project also illustrates how the concept of time works on an FPGA, so even if you don't have an FPGA to program, I recommend reading through this section.

Project #4: Debouncing a Switch

In Chapter 4, we programmed an FPGA to toggle an LED at the push of a button. However, there was a problem: pushing the button didn't consistently toggle the state of the LED. This is because any physical switch, including a push-button or toggle switch, is subject to *bouncing*, or rapid signal fluctuations that occur when the switch is toggled or flipped. Bouncing happens when the metal contacts inside the switch come together and move apart quickly before they have time to settle into the stable state. Figure 5-5 illustrates how this affects the switch's output signal.

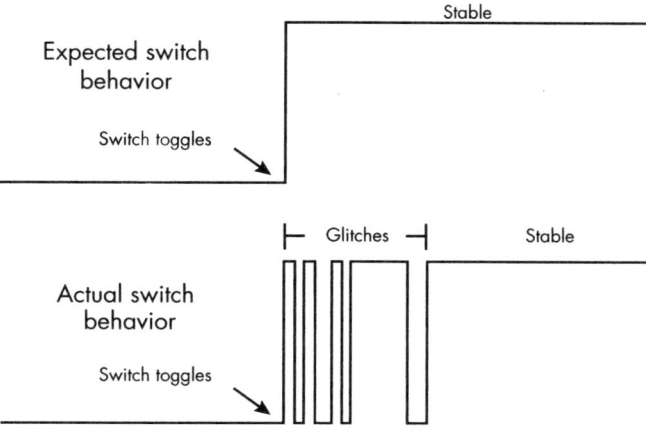

Figure 5-5: Bouncing in a mechanical switch

If you didn't know about bouncing, you would probably expect the switch to behave like the top half of Figure 5-5. The button is pressed, and the output immediately goes from low to high. However, in the real world bouncing creates glitches in the output signal, which show up as rapid low-to-high-to-low transitions of the output signal, before it finally stays high. Again, this is due to the mechanical switch contacts quickly coming together and moving apart before settling into a stable output state.

The code in our LED toggling project was looking for a single falling edge to indicate the press and release of the button, but due to the bouncing, the FPGA was seeing many falling edges per press/release. If it saw an odd number of falling edges during the bouncing of the switch, then the LED toggled successfully. If it saw an even number of falling edges, however, the LED didn't appear to change state, since each pair of falling edges effectively canceled each other out.

The number of bounces on a switch is somewhat random, so pushing the switch enough times got the LED to toggle successfully. Still, it would be better if the LED toggled as expected each time the switch is pressed and released. To make this happen, we need to add a *debounce filter* to the switch. That is, we need to program the FPGA to ignore the bounces. Figure 5-6 illustrates how this will work.

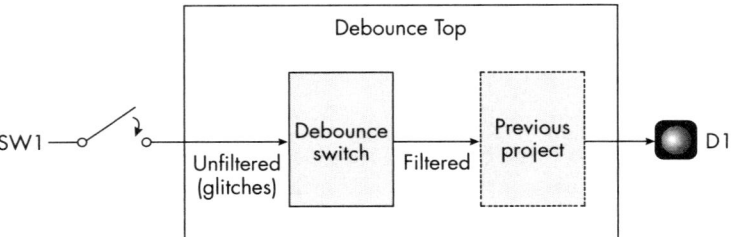

Figure 5-6: The Project #4 block diagram

We'll add a debounce filter to the code from the previous project to ensure that a single press of the button only toggles the LED once. The signal from the switch will pass through the debounce filter before going on to the LED-toggling logic we wrote in the last chapter.

We create a debounce filter by making sure that the input from the switch is stable for some amount of time before allowing the output driving the LED to change. We therefore need to have some concept of how much time has passed in our FPGA. However, introducing the notion of time into an FPGA design raises some interesting challenges.

Measuring Time on an FPGA

Time doesn't exist inherently in an FPGA. The FPGA doesn't automatically know if it's Saturday at 11:00 AM, or how to wait for 100 ms, for example. To be sure, there are parts of Verilog and VHDL code that refer to time. For example, we already saw how to use #10 in Verilog or wait for 10 ns; in VHDL to add 10 ns delays to our AND gate testbench. To give another example, in Verilog you can use $time to get the current time, while in VHDL the reserved word now gets a timestamp of the current time. However, while features like these will work perfectly well in simulation, they will 100 percent not work on your FPGA. They *aren't synthesizable*.

We've already talked about synthesis a few times. It's the part of the build process where the FPGA tool turns your Verilog or VHDL code into flip-flops, LUTs, and other components. Unfortunately, synthesis tools can't synthesize anything relating to time. It's just not possible. As a result, language constructs like $time and now are simply ignored or will create errors during synthesis. In Chapter 7, we'll look more closely at what features of VHDL and Verilog aren't able to be synthesized in an FPGA. For now, take it for granted that we can't use some of these built-in features relating to time.

If time doesn't exist in an FPGA, how can you keep track of how much time has passed for the purposes of debouncing a switch or one of the many other time-related tasks you may wish your FPGA to perform? The answer is to *count clock cycles*. If you know how many clock cycles have occurred, and you know the period of the clock, you'll know how much time has elapsed. Let's walk through an example.

Say you have a clock that oscillates at 25 MHz, and that the clock's period—the duration of a single cycle—is 40 ns. Given these specifications, how many clock cycles would it take for 400 ns to elapse? Answer: 10. And for 4,000 ns to elapse? Answer: 100. Simply divide the amount of time you want to wait by the period of the clock to get the number of clock cycles you need to count before that amount of time has elapsed. This technique is going to be critical in our debounce project.

Writing the Code

Let's look at how to implement the debounce filter. We'll start with the top-level module, which instantiates and links together two lower-level modules, one for debouncing the switch and the other for toggling the LED:

Verilog
```
module Debounce_Project_Top
  (input  i_Clk,
   input  i_Switch_1,
   output o_LED_1);

  wire w_Debounced_Switch;

❶ Debounce_Filter ❷ #(.DEBOUNCE_LIMIT(250000)) Debounce_Inst
  (.i_Clk(i_Clk),
   .i_Bouncy(i_Switch_1),
   .o_Debounced(w_Debounced_Switch));

❸ LED_Toggle_Project LED_Toggle_Inst
  (.i_Clk(i_Clk),
   .i_Switch_1(w_Debounced_Switch),
   .o_LED_1(o_LED_1));

endmodule
```

```
VHDL   library ieee;
       use ieee.std_logic_1164.all;

       entity Debounce_Project_Top is
         port (
           i_Clk        : in  std_logic;
           i_Switch_1   : in  std_logic;
           o_LED_1      : out std_logic
           );
       end entity Debounce_Project_Top;

       architecture RTL of Debounce_Project_Top is
         signal w_Debounced_Switch : std_logic;
       begin

     ❶ Debounce_Inst : entity work.Debounce_Filter
           generic map(
         ❷ DEBOUNCE_LIMIT => 250000)
           port map (
             i_Clk        => i_Clk,
             i_Bouncy     => i_Switch_1,
             o_Debounced  => w_Debounced_Switch);

     ❸ LED_Toggle_Inst : entity work.LED_Toggle_Project
           port map (
             i_Clk        => i_Clk,
             i_Switch_1   => w_Debounced_Switch,
             o_LED_1      => o_LED_1);

       end architecture RTL;
```

The code matches the block diagram in Figure 5-6. At the highest level we have Debounce_Project_Top, which instantiates two other modules. The first is the new debounce filter ❶, which we'll examine next. The second is the LED_Toggle_Project module that we created in the previous chapter ❸. It's worth taking a minute to follow the signals here. We can see the input signal i_Switch_1 going into the debounce filter. Out of that comes w_Debounced_Switch, which is the debounced version of this input. This is passed into the LED_Toggle _Project module. The output of that module is o_LED_1, which will be connected to the LED pin on your development board. Note that indicating the direction of your signals via their names, as we do here with the i_ and o_ prefixes, becomes very helpful as your designs get larger and incorporate more signals.

It's important to highlight the value of creating reusable modules when writing FPGA code. Rather than writing all the project's code from scratch, here we're able to reuse the LED_Toggle_Project module from the previous chapter and improve its functionality by interfacing it with another module. Another way to make modules reusable is to incorporate Verilog *parameters* or VHDL *generics*. These are variables within a module that you can override from higher-level code. We do this when we instantiate the Debounce_Filter module. Specifically, we override the module's parameter/generic called DEBOUNCE_LIMIT with the value 250000 ❷. As you'll see later, this value sets the number of clock cycles to wait while debouncing the switch. Coding it as a parameter/generic makes it easy to modify the value. In general, parameters (in Verilog) and generics (in VHDL) are a very useful way to keep code portable. They let you change the behavior of a module without having to actually modify the module's file.

Let's now examine the code of the debounce filter module:

Verilog

```verilog
module Debounce_Filter #(parameter DEBOUNCE_LIMIT = 20) (
   input   i_Clk,
   input   i_Bouncy,
   output o_Debounced);

❶ reg [$clog2(DEBOUNCE_LIMIT)-1:0] r_Count = 0;
   reg r_State = 1'b0;
   always @(posedge i_Clk)
   begin
   ❷ if (i_Bouncy !== r_State && r_Count < DEBOUNCE_LIMIT-1)
     begin
        r_Count <= r_Count + 1;
     end
   ❸ else if (r_Count == DEBOUNCE_LIMIT-1)
     begin
        r_State <= i_Bouncy;
        r_Count <= 0;
     end
     else
     begin
     ❹ r_Count <= 0;
     end
   end

❺ assign o_Debounced = r_State;
endmodule
```

VHDL

```vhdl
library ieee;
use ieee.std_logic_1164.all;
use ieee.numeric_std.all;

entity Debounce_Filter is
  generic (DEBOUNCE_LIMIT : integer := 20);
  port (
     i_Clk      : in  std_logic;
     i_Bouncy   : in  std_logic;
     o_Debounced : out std_logic
     );
end entity Debounce_Filter;

architecture RTL of Debounce_Filter is
❶ signal r_Count : integer range 0 to DEBOUNCE_LIMIT := 0;
   signal r_State : std_logic := '0';

begin

  process (i_Clk) is
```

```
    begin
      if rising_edge(i_Clk) then

    ❷ if (i_Bouncy /= r_State and r_Count < DEBOUNCE_LIMIT-1) then
         r_Count <= r_Count + 1;

    ❸ elsif r_Count = DEBOUNCE_LIMIT-1 then
         r_State <= i_Bouncy;
         r_Count <= 0;

      else
      ❹ r_Count <= 0;

      end if;
    end if;
  end process;

❺ o_Debounced <= r_State;
end architecture RTL;
```

The overall purpose of this module is to remove any bounces or glitches in the input (i_Bouncy) and create a stable output (o_Debounced). To do this, we check if the input and output are different. If they are, we know the input is changing, but we don't want to immediately update the output, since the switch might still be bouncing. Instead, we want to make sure that the input is stable for a long enough period of time before updating the output. Because the FPGA has no inherent concept of time, we implement the delay by counting clock cycles.

Let's say we want the input to be stable for 10 ms before we update the output. We need to count up to some number of clock cycles that represents 10 ms (or 10 million nanoseconds) of time passed. The Go Board, for example, has a clock period of 40 ns, so in this case we divide 10 million by 40 to get a delay of 250,000 clock cycles. This is the value we used for the DEBOUNCE_LIMIT parameter/generic in the top-level module Debounce_Project_Top. If your development board has a different clock period, you'll need to change DEBOUNCE_LIMIT accordingly.

The code used to create our clock cycle counter ❶ differs between the Verilog and VHDL versions. In Verilog, we use a common trick: the $clog2() built-in function (short for ceiling log base 2) determines the $\log_2$ of the number of clock cycles we want to count, rounded up. This tells us the number of binary digits needed to implement the counter. Thanks to the $clog2() function, we can dynamically size the r_Count register based on the input parameter, so if the input parameter changes (because your clock has a different period, or because you want to extend the wait time), the code will see this and synthesize r_Count to be as wide as it needs to be. This is better than hardcoding r_Count to some arbitrary limit, which could break when the code is reused.

With VHDL, we're able to achieve the same dynamic sizing in a simpler way, using the range keyword. This not only will size the variable correctly, but has an added benefit of creating a warning in your simulation if the value of r_Count ever goes beyond the integer range limit. The fact that the simulator can provide these types of warnings when running your testbench is another great reason for using simulations.

We implement the debounce filter using a series of if statements that are evaluated at each clock cycle. First we handle the case where the input is different from the output (meaning the input is changing) but r_Count is less than DEBOUNCE_LIMIT-1 ❷. This means we haven't yet waited the desired amount of time for the switch to stop bouncing, so we increment our clock cycle counter by 1. In this if statement, we're effectively waiting for some amount of time to pass to ensure the input is stable, before updating the output value.

Next we handle the case where the counter has reached its limit, so we know that we've waited the full 10 ms (or whatever length of time DEBOUNCE _LIMIT corresponds to) ❸. At this point, we can register the current value of the input (i_Bouncy) to r_State, whose value is in turn assigned to the output (o_Debounced) ❺. We also reset the counter to 0 to prepare for the next event. Finally, the else statement ❹ covers situations where the input and output have the same state. In this case, we reset the counter, since we have nothing to debounce here and we want our debounce filter to always be ready for the next event.

Creating the Testbench and Simulation

Now we'll create a testbench to exercise our project and make sure it works as expected. Recall that the testbench is what will instantiate our unit under test and simulate its inputs, while monitoring its outputs. In this case, we want the testbench to simulate the unstable input from a bouncing switch so we can confirm that the debounce filter is delaying the output until the switch has settled into a stable state. Here's the code:

Verilog
```
module Debounce_Filter_TB ();
   reg r_Clk = 1'b0, r_Bouncy = 1'b0;
❶ always #2 r_Clk <= !r_Clk;

❷ Debounce_Filter #(.DEBOUNCE_LIMIT(4)) UUT
   (.i_Clk(r_Clk),
    .i_Bouncy(r_Bouncy),
    .o_Debounced(w_Debounced));

❸ initial begin
     $dumpfile("dump.vcd"); $dumpvars;

     repeat(3) @(posedge r_Clk);
❹ r_Bouncy <= 1'b1; // toggle state of input pin

     @(posedge r_Clk);
❺ r_Bouncy <= 1'b0; // simulate a glitch/bounce of switch

     @(posedge r_Clk);
❻ r_Bouncy <= 1'b1; // bounce goes away

     repeat(6) @(posedge r_Clk);
     $display("Test Complete");
```

```
        $finish();
    end
endmodule
```

VHDL
```
library ieee;
use ieee.std_logic_1164.all;
use std.env.finish;
entity Debounce_Filter_TB is
end entity Debounce_Filter_TB;

architecture test of Debounce_Filter_TB is
  signal r_Clk, r_Bouncy, w_Debounced : std_logic := '0';
begin
❶ r_Clk <= not r_Clk after 2 ns;

❷ UUT : entity work.Debounce_Filter
    generic map (DEBOUNCE_LIMIT => 4)
    port map (
      i_Clk       => r_Clk,
      i_Bouncy    => r_Bouncy,
      o_Debounced => w_Debounced);

❸ process is
  begin
    wait for 10 ns;
  ❹ r_Bouncy <= '1';  -- toggle state of input pin

    wait until rising_edge(r_Clk);
  ❺ r_Bouncy <= '0';  -- simulate a glitch/bounce of switch

    wait until rising_edge(r_Clk);
  ❻ r_Bouncy <= '1';  -- bounce goes away

    wait for 24 ns;
    finish;  -- need VHDL-2008
  end process;
end test;
```

Unlike our AND gate testbench, this testbench must provide a clock signal to the UUT, along with the other inputs. We create the clock signal with a simple trick ❶: we repeatedly invert a signal after a fixed amount of time to generate a 50 percent duty cycle signal that will toggle for the duration of the testbench execution. The signal inverts every 2 ns, for a clock period of 4 ns per cycle. This is much faster than the actual clock period on a typical FPGA development board, but for the purposes of this simulation, that's okay.

When we instantiate the UUT ❷, we override DEBOUNCE_LIMIT with the value 4. This means our debounce filter will only look for four clock cycles of stability before it deems the output debounced. In a real FPGA, this would be a very short amount of time (less than 1 microsecond), probably not long enough to actually fix the problem. However, keep in mind the purpose of this testbench: we want to make sure that our FPGA logic works as intended. That logic is functionally the same whether we're waiting 4 clock cycles or 250,000 clock cycles. Using the much smaller number will make for a quicker simulation and an easier-to-evaluate waveform, while still giving us realistic feedback about whether or not the design works. Shortening counters is a handy trick to remember for large designs: a simulation of such a design could take many minutes to run, but using smaller limits for counters will make the simulation run faster, allowing you to debug your code more quickly. Once your design is fully debugged and verified, you can update the simulation with your actual expected counter lengths to validate your actual design. This means you'll only have to endure the longer simulation time once, after any issues with the code have already been resolved using the shortened simulation.

Next, we start to provide stimulus to the UUT ❸. With synchronous designs, we want to ensure that our input signals to the UUT are synchronous to the clock. We therefore set up the code to change the stimulus on the rising edge of the testbench clock. Otherwise, we might be introducing strange timing effects that would not exist in a real FPGA design. (Remember that all of the flip-flops in your UUT will be using the rising edge of the clock, so your testbench stimulus should also be reacting to the rising edge of the clock.)

When the test starts, the input is low. After a short time, the input goes high for a single clock cycle ❹, then low again ❺, to simulate a bouncing-induced glitch. We want to make sure that the debounced output of this module doesn't react to this glitch. Later in the test, we drive the input back high again and leave it there ❻. This time we want the output to match the input, but only after the debounce filter has counted out four clock cycles.

After running this testbench code in EDA Playground (or whichever simulator you prefer), you should get a waveform that looks something like that shown in Figure 5-7.

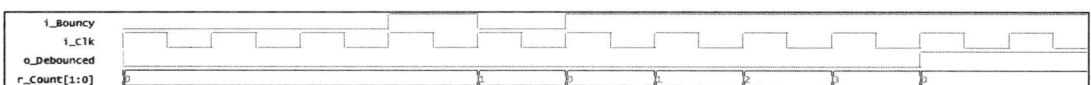

Figure 5-7: The debounce simulation waveform

The waveform shows that the output o_Debounced stays low when i_Bouncy goes high for only one clock cycle. Then, toward the end of the simulation, we see the output go high to match the input, but only after the input has been high for four clock cycles. The debounce filter works!

While the testbench we wrote is better than no test at all, it could certainly be improved. For example, we don't check what happens when the input goes low again, to make sure the output responds correctly. Additionally, we might want to check that a higher DEBOUNCE_LIMIT value than 4 doesn't cause any issues. Setting up multiple tests to stress the bulk of your design, and in particular any corner cases, is good test design practice.

Building and Programming the FPGA

Having simulated the design, we now have some confidence that if we were to go ahead and program the FPGA, it would likely work as intended. Let's try it out! Create a new project inside iCEcube2, and add the following modules to the project: Debounce_Filter, Debounce_Project_Top, and LED_Toggle_Project. Make sure that you've also included the clock constraints file, as well as the physical constraints file.

When everything's ready, build the project. Then double-check for errors, and check your utilization reports. The synthesis report after building the FPGA will look something like this:

```
--snip--
Register bits not including I/Os:    21 (1%)
Total load per clock:
    i_Clk: 1

Mapping Summary:
Total  LUTs: 31 (2%)
--snip--
```

From this report, we can see that we're using more LUTs and flip-flops than we did for Project #3. This makes sense; the debounce filter accounts for these extra resources. Still, the FPGA has plenty of resources to spare.

Go ahead and program your FPGA, then try pushing the button to turn the LED on and off. You should notice that the LED is now toggling consistently with each press of the button. We've successfully filtered out the bounces from the switch.

As you've seen in this project, simulations are invaluable for building confidence in a design and debugging issues in your Verilog and VHDL. However, even in this relatively simple example, you may have noticed a drawback: examining waveforms to determine if a design is working can be tedious, especially if you have to keep changing the design and rerunning the simulation. It would be much more convenient if the testbench could simply tell you whether the simulation has worked, without you having to study the waveform. As we'll explore next, it's possible to write testbenches that offer exactly this capability.

Self-Checking Testbenches

A *self-checking testbench* is a testbench that you program to verify your UUT is working as intended, without having to manually inspect the output. The self-checking testbench will run a series of steps and let you know if any have failed, at which point you can inspect the failure and fix it. This saves you from having to visually examine the waveforms generated during simulation to determine whether your design has worked as expected. It takes a bit more effort to set up your testbench to be self-checking, but it's almost always worth the time spent.

When setting up a self-checking testbench, your goal is to inject many different test cases into your UUT, then monitor the outputs and check, or *assert*, that they are what you expect. *Assertions* are statements about what value a signal will have at a particular moment in the simulation, and they're probably the most critical part of a self-checking testbench. Often, a self-checking testbench will have hundreds of assertions, with each one imparting a little more confidence that the design is correct.

Self-checking testbenches are particularly useful if you're adding a new feature to some old code. It might be something that you haven't looked at in years, and suddenly you need to try to remember (or learn, if someone else wrote it) how it works. From experience, I can tell you that starting with a testbench that has many checks is a huge benefit. You'll be able to open the simulation, see all of the assertions in the self-checking testbench, and make sure everything in the old code still works. Then you can add your new code and add new tests for it. Once all the old *and* new tests are passing, you can have high confidence that your new code is performing as expected—and equally importantly, that *you haven't broken any old code.*

To illustrate how self-checking testbenches work, let's return to the simple testbench we wrote for our AND gate project earlier in the chapter. The following Verilog and VHDL code takes the original testbench we wrote and adds some assertion checks within it. These assertions will automatically run to verify that the actual output is in the expected state. The new code is shown in bold:

Verilog
```
--snip--
  initial
    begin
      $dumpfile("dump.vcd"); $dumpvars;
      r_In1 <= 1'b0;
      r_In2 <= 1'b0;
      #10;
    ❶ assert (w_Out == 1'b0);

      r_In1 <= 1'b0;
      r_In2 <= 1'b1;
      #10;
      assert (w_Out == 1'b0);
--snip--
```

VHDL
```
--snip--
process is
  begin
    r_In1 <= '0';
    r_In2 <= '0';
    wait for 10 ns;
  ❶ assert (w_Out = '0') severity failure;

    r_In1 <= '0';
    r_In2 <= '1';
    wait for 10 ns;
    assert (w_Out = '0') severity failure;
--snip--
```

In this excerpt from the testbench, we've added two checks. We use the assert keyword ❶ to first confirm that the output is low when both inputs are low, then that it's low when one input is low and the other is high. The assert keyword only exists in SystemVerilog, not regular Verilog. This is an example of how SystemVerilog has improved features for testbenches. VHDL, meanwhile, has assert built into it, and the severity can be note, warning, or failure, depending on the level of assertion that you want to check for. Each has a different escalation, so you can filter them out in your report. In this case, we've chosen failure, since we definitely wouldn't want an AND gate output high when the inputs are low.

If this assertion evaluates to true, then the simulation moves on. However, if something goes wrong and the assertion fails, you'll see output printed to the screen. In Verilog, you'd see something like this:

```
# ASSERT: Error: ASRT_0301 testbench.sv(20): Immediate assert
condition (w_Out==1'b1) FAILED at time: 10ns, scope: And_Gate_TB
```

In VHDL, here is what the failure message would look like:

```
# ** Failure: Assertion violation.
#    Time: 10 ns Iteration: 0  Process: /and_gate_tb/line__22
File: testbench.vhd
```

This is very helpful! Not only do we know that the testbench failed, but we know that it failed exactly 10 ns into the simulation, which allows us to immediately locate the failure in the waveform viewer. We also know the exact line of code that caused the failure: line 20 in Verilog or line 22 in VHDL. These pieces of information make it easier to investigate the problem, understand the cause, and fix it. I recommend adding assertions into your tests wherever possible.

The self-checking testbench is an area where SystemVerilog really shines. Many of the added features beyond what regular Verilog offers are geared toward writing better testbenches. For example, SystemVerilog provides the ability for you to verify sequences of events. This can be handy for analyzing interactions between different signals, to make sure they happen correctly (that is, first one thing happens, then on the next clock cycle, another thing happens). SystemVerilog also provides classes, allowing you to use object-oriented programming techniques to streamline your testbench code. Other SystemVerilog features allow you to randomly inject data into your designs, making your tests much more comprehensive and robust. The details of these features are beyond the scope of this book, but as you start writing more testbenches—particularly self-checking testbenches—I encourage you to learn more about SystemVerilog.

Initial Signal Conditions

By default, if a signal isn't assigned an initial condition, then it will show up in an unknown state when you start a simulation. This is often represented by a red signal and an X in the waveform viewer. The simulator is telling you that it doesn't know how to treat the signal when the testbench is first running. Should it be a 0 or a 1? The simulator doesn't know.

There are two methods for assigning a default state to your signals, so that they start in a known state. One method is to use resets. As we discussed back in Chapter 4, a reset assigns an initial default value to a flip-flop. Driving the reset input at the start of a simulation will set the signals to known states to begin the test. This will work for all signals that are assigned a reset condition.

The other way we can set signals to an initial state is to use the initialization feature in both Verilog and VHDL. This is particularly useful for simulation purposes. It's as simple as assigning a signal to a value after it's created. In Verilog, for example, reg r_State = 1'b0; initializes the r_State signal to 0. In VHDL, signal r_State : std_logic := '0'; does the same. You can use any state that the signal can validly be set to as an initialization value.

Initial signal assignments are only synthesizable for some FPGAs, since not all FPGAs can load an initial state into their flip-flops when they boot up after being programmed. Because this feature isn't available for all FPGAs, I generally don't recommend relying on it. A better, more portable solution is to use resets to set signals to some default value. Resets are widely supported across all FPGA manufacturers, so your code will be more portable if you need to change FPGAs.

On-FPGA Debugging

Early in this chapter, I told you that once you're on hardware, you're looking at a black box. You can see inputs and outputs, but you can't see what's going on internally. This isn't entirely true. There *is* a way to do some limited on-FPGA debugging. However, this method has significant drawbacks and should only be used sparingly, if at all.

On-FPGA debugging is achieved by adding a *logic analyzer,* a tool that shows the state (high or low) of many digital signals at once, inside your FPGA. This allows you to monitor the FPGA's internal signals in real time. By looking at these signals, you can debug issues and see where data isn't behaving as expected.

Each of the major FPGA companies has a unique product within its suite of tools that creates a logic analyzer inside the FPGA. AMD has a feature called Integrated Logic Analyzer (ILA), Intel has Signal Tap, and Lattice has Reveal. They all work basically the same way: they take part of your FPGA's resources and turn those resources into a logic analyzer. You run your FPGA code, the logic analyzer "sniffs" the data, and the results are presented on your computer screen so you can debug your design.

There are several problems with this process, however. The first issue is that it's extremely time-consuming. If you want to add a logic analyzer to your FPGA, you need to rebuild and reprogram the entire design. You also need to decide ahead of time what signals you're interested in monitoring with the logic analyzer, as you likely won't have enough resources on your

FPGA to look at everything. If you want to change what you're looking at while the FPGA is running, too bad! You'll have to rebuild the entire FPGA from scratch and start the process all over again. A simulation, on the other hand, can easily see the state of *all* the signals on your FPGA; you don't have to pick and choose.

An additional problem with on-FPGA debugging is that adding a logic analyzer is basically throwaway effort. Once you find and fix your one problem, you don't need the debug tool anymore. In fact, since it uses your FPGA's resources (which are a limited commodity), you may not want to keep it in your design. You can save and rerun a simulation, but a logic analyzer is a one-and-done debugging effort.

The final and perhaps worst problem is that when you add a logic analyzer to your FPGA, you're changing the FPGA's design, which can have unintended consequences. Issues that are subject to small timing variations might be fixed by the very act of adding the logic analyzer, or new issues might be created. If you're trying to use the logic analyzer to debug a race condition inside your FPGA, for example, the design changes that result from adding in the logic analyzer might actually make the race condition go away. Scientists refer to this as the *observer effect*, where a phenomenon is changed by the act of investigating it.

This isn't to say that these on-FPGA debuggers are entirely useless. They're helpful when you're trying to investigate a situation that's difficult to simulate. For example, say some external interface to your FPGA is causing problems, but those problems are only occurring on your hardware while the simulation is working fine. At that point you might want to fire up a logic analyzer and try to see why your simulation is different from real life. Once you figure it out, you should strive to make your simulation as realistic as possible, adding to it the failure mode that you identified with the real-world test.

These tools have saved me a couple of times in my career, but in general I try to avoid them if possible.

Verification

Verification is the process of ensuring that an FPGA or ASIC design is working as intended. It's an exhaustive process that goes well beyond writing a few testbenches and running a simulation—so much so that there are people called verification engineers who perform verification full time. The complete details of how verification works are beyond the scope of this book. In fact, there are entire books dedicated to the subject. This section simply introduces the topic so you're aware of the key role verification can play in real-world FPGA and ASIC design.

Consider a device like a DVD player. What happens if a DVD is playing, then the user pauses playback, ejects the DVD, and presses the fast-forward button? Does the code handle that sequence of events correctly? Or does the unexpected fast-forward command lock up the processor in a strange state? The verification engineer must test all of these corner cases to ensure the design won't make a mistake in handling some strange situation.

Back in Chapter 1, I mentioned that making an ASIC is an incredibly expensive and time-consuming process. Once the ASIC is fabricated at the foundry, you have to cut a big check. If there are critical bugs in the design and the ASIC doesn't work as intended, then you've just lost all that money, and you'll need to rerun the ASIC fabrication process again. It's the job of a verification engineer to ensure the design is correct up front, since finding and fixing bugs later is incredibly expensive.

Squashing bugs is great, but that's only half the benefit. Another major goal of verification is to ensure the design is performing as intended. If you're handed a specification of how an ASIC is supposed to perform, there might be ambiguities or missing information. Usually, one or more designers will design to the specification, and one or more separate verification engineers will simultaneously verify that the design is meeting the specification. If any discrepancies arise, the two teams can get together and update the specification so that everyone is clear on the intent.

Most verification engineers take advantage of the extra features built into SystemVerilog to thoroughly test a design. Self-checking testbenches are absolutely a must. It's helpful to exercise the design randomly as well, so there are blocks of code that can inject random test cases into the design and ensure it's working as intended.

Verifying code like this is no small feat. Often it's more expensive and time-consuming to verify a design is working correctly than to create the design itself! For this reason, unlike with ASICs, not many FPGA designs go through a dedicated verification process. In most cases, it's just too expensive. Remember that FPGA stands for *field programmable* gate array, so if a few bugs are allowed to slip through, the device can always be updated in the field, or in the hands of a customer.

Usually, only FPGA designs that demand very high reliability or simply cannot be updated in the field go through verification. For example, some FPGAs are *one-time programmable (OTP)*, meaning they can only be programmed once; afterward, the functionality is locked in and cannot be changed. Some applications in outer space utilize these OTP FPGAs, since they're more resistant to radiation. Additionally, OTP FPGAs are considered less susceptible to reverse engineering, so they're preferable for high-security applications. OTP FPGA designs often require verification; however, this isn't the norm for typical FPGA designs.

For our purposes, testbenches are sufficient to find bugs in the FPGA design, but for ASICs or FPGAs that require it, verification is critically important.

Summary

In this chapter, you've learned about simulating your FPGA code and seen how running simulations is time well spent. In contrast to debugging on hardware, simulation lets you visualize all the signals in your design, observe how they interact, and create stimuli to stress your code and see how it responds. You practiced writing testbenches, or code that instantiates your UUT, injects it with sample input, and monitors the output. You saw how looking at the waveforms generated during simulation is a great way to see if your design is working, but better still is adding tests that make your testbench self-checking. Less debugging on hardware, more beautiful simulations: that's what makes a happy FPGA designer.

6

COMMON FPGA MODULES

Working with an FPGA can feel like building with LEGO: you have a limited variety of small bricks at your disposal, but by stacking them elegantly, you can create amazingly complex designs. At the lowest level, you're working with LUTs and flip-flops. At a slightly higher level, there are several basic building blocks that appear over and over again in FPGA designs, including multiplexers and demultiplexers, shift registers, and first in, first out (FIFO) and other types of memory.

Each of these elements is very common. In fact, it's likely that one or more of them will be used in every single FPGA project you'll ever work on. In this chapter, I'll show you how these basic building blocks work and how to implement them with Verilog and VHDL. For each of these common elements, you'll create a self-contained module that you can reuse anytime you need that element in an FPGA design. This will reinforce your FPGA programming knowledge and give you a solid foundation for your own projects.

Multiplexers and Demultiplexers

Multiplexers and *demultiplexers* are circuit components that allow you to select between two or more things. In the case of a multiplexer (sometimes spelled *multiplexor*, and often shortened to *mux*), you have multiple input signals, and you select which of them is sent to a single output. A demultiplexer (*demux* for short) is the opposite: you have a single input signal, and you select which of multiple outputs it should go to.

Multiplexers and demultiplexers have many applications. For example, a mux could be used to select which speed to run a fan at: the low-med-high switch might be acting as a mux to control which setting is sent to the fan controller. A demux could work with a switch to select which of four LEDs to illuminate: only one LED will be illuminated at a time, but you'll be able to specify which one is illuminated.

Muxes and demuxes are classified based on how many inputs and outputs they have. For example, a 4-1 (pronounced *four-to-one*) mux has four inputs and one output. Conversely, a 1-4 (pronounced *one-to-four*) demux has one input and four outputs. You can design muxes to have any number of inputs, depending on the requirements of your circuit: you can have a 2-1 mux, a 3-1 mux, an 8-1 mux, a 13-1 mux, or whatever you want. Likewise, you can design a demux with however many outputs you need.

Implementing a Multiplexer

Let's consider how to create a multiplexer on an FPGA. Specifically, we'll look at creating a 4-1 mux, but you can apply the same logic to a mux with any number of inputs. Figure 6-1 shows a block diagram of a 4-1 mux.

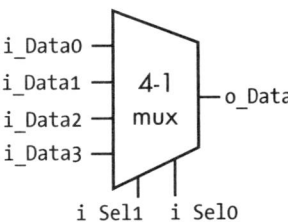

Figure 6-1: A 4-1 multiplexer
(mux)

Our multiplexer has four inputs on the left: i_Data0, i_Data1, i_Data2, and i_Data3. On the right is the single output, called o_Data. At the bottom are two additional inputs, labeled i_Sel1 and i_Sel0. *Sel* is short for *select*. These selector inputs choose which of the four data inputs is passed to the output. The truth table in Table 6-1 shows how i_Sel1 and i_Sel0 work together to determine the mux's output.

Table 6-1: Truth Table for a 4-1 Mux

i_Sel1	i_Sel0	o_Data
0	0	i_Data0
0	1	i_Data1
1	0	i_Data2
1	1	i_Data3

Looking at Table 6-1, we can see that i_Data0 is connected to the output when i_Sel1 and i_Sel0 are both 0. The output gets i_Data1 when i_Sel1 is 0 and i_Sel0 is 1, it gets i_Data2 when i_Sel1 is 1 and i_Sel0 is 0, and it gets i_Data3 when both selectors are 1.

NOTE *Because muxes serve to select which inputs go to which outputs, they're often called* selectors. *In fact,* select *is a reserved word in VHDL that can be used to generate muxes.*

Implementing this truth table in Verilog or VHDL is simply a matter of evaluating i_Sel1 and i_Sel0 and assigning the appropriate data input to the output. The following listing shows how it's done (I've omitted the signal definitions to focus on the actual mux code, but more context can be found in the book's GitHub repository, at *https://github.com/nandland/getting-started -with-fpgas*):

Verilog
```
assign o_Data = !i_Sel1 & !i_Sel0 ? i_Data0 :
                !i_Sel1 &  i_Sel0 ? i_Data1 :
                 i_Sel1 & !i_Sel0 ? i_Data2 : i_Data3;
```

VHDL
```
o_Data <= i_Data0 when i_Sel1 = '0' and i_Sel0 = '0' else
          i_Data1 when i_Sel1 = '0' and i_Sel0 = '1' else
          i_Data2 when i_Sel1 = '1' and i_Sel0 = '0' else
          i_Data3;
```

The Verilog version uses the conditional (or ternary) operator, represented by a question mark (?). This is shorthand for writing conditional expressions without using if...else statements. The operator works by first evaluating the condition before the question mark (for example, !i_Sel1 & !i_Sel0). If the condition is true, the expression selects the condition before the colon. If the condition is false, it selects the condition after the colon. Here, we've chained several ? operators together to handle each possible combination of the two selector inputs.

In the VHDL version, we accomplish the same thing by chaining several when/else statements. Since the VHDL version uses more spelled-out keywords, it's a bit more readable, but the Verilog is more concise. In both the Verilog and VHDL versions, the chain of logical checks gets evaluated until one evaluates as true. If none of them are true, then we use the last assignment in the chain.

Implementing a Demultiplexer

For a 1-4 demux, the block diagram looks like a mirrored version of a 4-1 mux, as you can see in Figure 6-2.

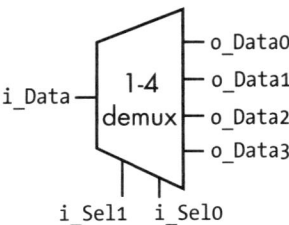

Figure 6-2: A 1-4 demultiplexer (demux)

This demux takes a single data input on the left (i_Data) and selects which output to connect it to. The demux is doing a 1-4 selection, so two input selectors are required to select between the four possible outputs. The truth table in Table 6-2 shows all the possible combinations.

Table 6-2: Truth Table for a 1-4 Demux

i_Sel1	i_Sel0	o_Data3	o_Data2	o_Data1	o_Data0
0	0	0	0	0	i_Data
0	1	0	0	i_Data	0
1	0	0	i_Data	0	0
1	1	i_Data	0	0	0

Looking at the table, we can see that i_Data is connected to one of the four outputs at a time, as determined by the i_Sel1 and i_Sel0 selector inputs. When i_Sel1 and i_Sel0 are both 0, o_Data0 gets i_Data; otherwise it gets 0. When i_Sel1 is 0 and i_Sel0 is 1, o_Data1 gets i_Data; otherwise it gets 0. When i_Sel1 is 1 and i_Sel0 is 0, o_Data2 gets i_Data; otherwise it gets 0. Finally, when i_Sel1 is 1 and i_Sel0 is 1, o_Data3 gets i_Data; otherwise it gets 0. Let's see how we can implement this truth table in Verilog and VHDL:

Verilog

```
module Demux_1_To_4
   (input  i_Data,
    input  i_Sel1,
    input  i_Sel0,
    output o_Data0,
    output o_Data1,
    output o_Data2,
    output o_Data3);

❶ assign o_Data0 = !i_Sel1 & !i_Sel0 ? i_Data : 1'b0;
   assign o_Data1 = !i_Sel1 &  i_Sel0 ? i_Data : 1'b0;
```

```
    assign o_Data2 = i_Sel1 & !i_Sel0 ? i_Data : 1'b0;
    assign o_Data3 = i_Sel1 &  i_Sel0 ? i_Data : 1'b0;

    endmodule
```

VHDL
```
library ieee;
use ieee.std_logic_1164.all;

entity Demux_1_To_4 is
  port (
    i_Data  : in  std_logic;
    i_Sel0  : in  std_logic;
    i_Sel1  : in  std_logic;
    o_Data0 : out std_logic;
    o_Data1 : out std_logic;
    o_Data2 : out std_logic;
    o_Data3 : out std_logic);
end entity Demux_1_To_4;
architecture RTL of Demux_1_To_4 is
begin

❶ o_Data0 <= i_Data when i_Sel1 = '0' and i_Sel0 = '0' else '0';
  o_Data1 <= i_Data when i_Sel1 = '0' and i_Sel0 = '1' else '0';
  o_Data2 <= i_Data when i_Sel1 = '1' and i_Sel0 = '0' else '0';
  o_Data3 <= i_Data when i_Sel1 = '1' and i_Sel0 = '1' else '0';
end architecture RTL;
```

Notice how each output in this code is set independently. The input i_Data can only be assigned to a single output at a time. For example, we assign it to the first output, o_Data0, when both selector inputs are 0 ❶. When an output isn't wired to the input data, then it's just set to 0 to disable it.

In practice, since a mux or demux can be created with just a few lines of code, it's unlikely that you'd ever create a module to instantiate a single multiplexer or demultiplexer. Generally, you'll be better off just putting the code that builds the mux or demux directly into the module where it's needed. However, multiplexers and demultiplexers are incredibly common circuit design elements, so it's important to understand how to implement them. Next, we'll look at another common component: the shift register.

The Shift Register

A *shift register* is a series of flip-flops where the output of one flip-flop is connected to the input of the next. We looked at a shift register back in Chapter 4 when we talked about a chain of flip-flops, but to keep things simple I didn't introduce the term at the time. To review, Figure 6-3 shows a chain of four flip-flops, which we can now call a *4-bit shift register*. As discussed in Chapter 4, each additional flip-flop in the chain adds a single clock cycle of delay to the output.

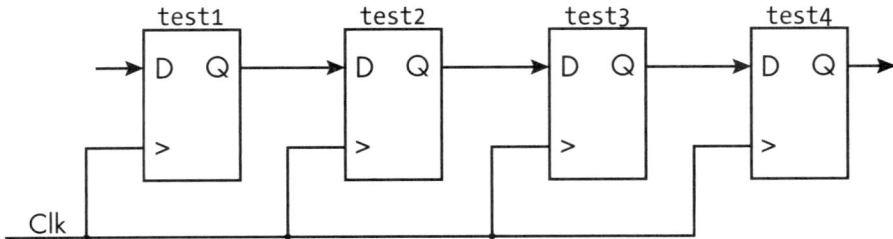

Figure 6-3: A shift register

Shift registers have many uses. For example, they can delay data for some fixed number of clock cycles, convert data from serial to parallel or from parallel to serial, or create a linear feedback shift register. We'll look at examples of each of these applications in this section.

Delaying Data

Creating delay in an FPGA is the most common application of a shift register. The delay is often used to align data in time. For example, when you send input data through a math operation, it might take a few clock cycles to produce a result. If you need to align the output result with the original input data, then the original input data needs to be delayed by the number of clock cycles that the math operation will take to perform.

As we've seen, a shift register is simply a chain of flip-flops, and the number of flip-flops in the chain dictates how many clock cycles it will take for the data on the input to propagate to the output. With that in mind, this code will create a shift register that generates a four-clock-cycle delay on some input data:

Verilog
```
❶ reg [3:0] r_Shift;
always @ (posedge i_Clk)
  begin
  ❷ r_Shift[0]   <= i_Data_To_Delay;
  ❸ r_Shift[3:1] <= r_Shift[2:0];
  end
```

VHDL
```
❶ signal r_Shift : std_logic_vector(3 downto 0);
process (i_Clk)
begin
  if rising_edge(i_Clk) then
  ❷ r_Shift(0)          <= i_Data_To_Delay;
  ❸ r_Shift(3 downto 1) <= r_Shift(2 downto 0);
  end if;
end process;
```

Here we create a shift register called r_Shift, which will be four flip-flops in length ❶. Remember, the r_ in the name is a clue that the signal will consist of flip-flops and be assigned within a clocked always block (in Verilog) or process block (in VHDL). We load up the first flip-flop in the chain (position 0) with i_Data_To_Delay, the input signal ❷. Then we use a trick to create the remaining three flip-flop assignments in a single line of code, rather than three: we take the values on flip-flops 0 through 2 and assign them to flip-flops 1 through 3 ❸. This way, the data that was on the first flip-flop in the chain is shifted to the second flip-flop, the data on the second flip-flop is shifted to the third, and so on. If you wanted, you could break this step down into its individual operations, like so:

```
r_Shift[3] <= r_Shift[2];
r_Shift[2] <= r_Shift[1];
r_Shift[1] <= r_Shift[0];
```

This example shows the Verilog version. For VHDL, replace the square brackets with parentheses.

Writing out each assignment individually demonstrates more explicitly how the data moves through the shift register one bit at a time, but both methods will work the same way. Now we can use bit position 3 of r_Shift for our purposes, as this is the flip-flop that represents a four-clock-cycle delay of the input data i_Data_To_Delay. If we needed a three-clock-cycle delay instead, we could use the data at bit position 2, or we could add more flip-flops to the chain to create a longer delay.

Converting Between Serial and Parallel Data

Converting from serial data to parallel data and vice versa is another common use of a shift register. You might need to do this when communicating with off-chip interfaces that transmit and receive data serially. One specific example is interfacing with a *universal asynchronous receiver-transmitter (UART)*. This is a device that transmits bytes of data by breaking them into individual bits, which are then reconstituted into bytes on the receiving end. When the data is sent, it is converted from parallel to serial: the eight parallel bits of data in a byte are sent serially, one after the other. When the data is received, it's converted back from serial (individual bits) to parallel (a complete byte).

UARTs are widely used to send and receive data between devices because they're simple and effective, and they're a perfect application for a shift register. An eight-bit shift register can send a byte of data by reading it out, one flip-flop at a time, or it can receive a byte of data by shifting the bits through the chain of flip-flops, one bit after the other. For example, say we want to send and receive ASCII-encoded characters, each of which can be represented within a single byte of data. First, let's look at the receiving end of the UART. Each line in Table 6-3 represents the receipt of a single bit of data. The column on the right shows how the complete byte is built up by shifting the bits through a shift register.

Table 6-3: Receiving a Byte of Data Through a UART

Bit index	Received bit	Byte contents
0	1	1
1	1	11
2	0	011
3	1	1011
4	0	01011
5	0	001011
6	1	1001011
7	0	01001011
		ASCII=0x4B='K'

UARTs normally receive data starting with the least significant (rightmost) bit. The first bit received is shifted through the shift register from the most significant (leftmost) position to the least significant position as more bits come in. Let's walk through how this works.

On the first line of the table, we've received the first bit, which has a value of 1. We place it into the most significant bit position, the first flip-flop in the shift register. When we receive the second bit, which is also a 1, we shift the existing bit to the right, and put the new received bit in the most significant bit position. The third bit we receive is a 0. Once again, we place it into the most significant position, and the rest of the bits are shifted right. Once we've received all eight bits, the shift register is full, with the last bit placed into the most significant bit position and the first bit placed in the least significant position. At this point, the byte is complete. In our example, we've received 01001011, which is equivalent to 0x4B (meaning 4B in hexadecimal), the ASCII encoding for the letter K. By receiving the data one bit at a time and shifting the received bits to the right with a shift register, we converted serial data to parallel data.

Now let's look at the transmit side of a UART. Table 6-4 shows how to transmit the byte 00110111, or 0x37, which is the digit 7 in ASCII.

Table 6-4: Transmitting a Byte of Data Through a UART

Bit index	Byte contents	Transmitted bit
	ASCII=0x37='7'	
0	00110111	1
1	0011011	1
2	001101	1
3	00110	0
4	0011	1
5	001	1
6	00	0
7	0	0

In this case, we start with the entire byte of data loaded in an 8-bit shift register. Again, a UART transmits from least significant bit to most significant bit, so here we send out the rightmost bit and shift the entire byte to the right with each step. By using a shift register to send out one bit at a time and shift the remaining bits to the right, we're converting parallel data to serial data.

Creating a Linear Feedback Shift Register

The last common application of the shift register is to create a *linear feedback shift register (LFSR)*. This is a shift register where certain flip-flops in the chain are tapped into and used as input for either an XOR or an XNOR gate (we'll be using XNOR). The output of this gate is then fed back into the beginning of the shift register, hence the word *feedback* in the name. *Linear* comes from the fact that this arrangement produces an input bit that's a linear function of the LFSR's previous state. Figure 6-4 shows an example of a 3-bit LFSR, but keep in mind that LFSRs can have any number of bits.

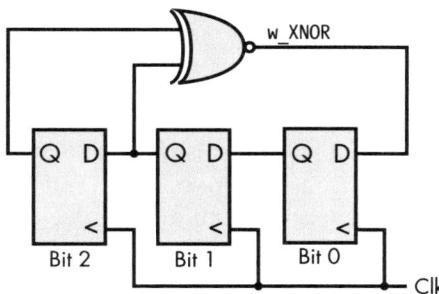

Figure 6-4: A 3-bit LFSR

This LFSR features three chained flip-flops, representing bits 0 through 2 of a shift register. The outputs of the bit 1 flip-flop and the bit 2 flip-flop are passed through an XNOR gate, and the output of the gate is sent to the input of the first bit in the shift register. The value of the LFSR at any given clock cycle is the value of the three flip-flop outputs.

NOTE   *The flip-flops in Figure 6-4 are drawn backward compared to how we've usually seen them, with input D on the right and output Q on the left. I drew them this way so the least significant bit (bit 0) would appear on the right, to match how we write numbers, but there's nothing special here; these are the same flip-flops that we all know and love, just mirrored.*

When an LFSR is running, the pattern generated by the individual flip-flops is *pseudorandom*, meaning it's close to but not completely random. It's only pseudorandom because from any state of the LFSR pattern, you can predict the next state. Table 6-5 shows what happens when the 3-bit LFSR is initialized to zero, then the clock starts toggling.

Table 6-5: Pseudorandom Output of 3-Bit LFSR

Clock cycle	LFSR data (binary)	LFSR data (decimal)
0	000	0
1	001	1
2	011	3
3	110	6
4	101	5
5	010	2
6	100	4
7	000	0
8	001	1
9	011	3
10	110	6
. . .	. . .	. . .

The LFSR goes from 000 to 001 on the first clock cycle. This makes sense, because the XNOR of bit 2 (0) and bit 1 (0) is 1, which gets written into bit 0. On the next clock cycle, the LFSR goes from 001 to 011. Once again we've taken the XNOR of bit 2 (0) and bit 1 (0), giving us a new bit 0 value of 1. Meanwhile, the old bit 0 value (1) has shifted to bit 1. Following the rest of the values in the table, they seem relatively random—pseudorandom, even!

Notice that the table repeats itself on the seventh clock cycle, so there are seven unique values that the 3-bit LFSR can have: 000, 001, 010, 011, 100, 101, and 110. It can never have a value of 111. If you're wondering why, consider what would happen if this value arose. At the next clock cycle, the new bit 0 would be the XNOR of 1 and 1, which is 1, while the other bits would shift over, giving us 111 again. The LFSR would be stuck on 111 forever, so it would effectively stop running! As a rule, for an LFSR that's N bit positions long, the maximum number of clock cycles that the LFSR takes to run through all combinations is $2^N - 1$. For 3 bits, it's $2^3 - 1 = 7$; for 4 bits, it's $2^4 - 1 = 15$; and so on.

Because of their pseudorandomness, LFSRs have many applications. They can function as low-utilization counters, test pattern generators, data scramblers, or be used in cryptography. The LFSR is lightweight, so these kinds of mathematical operations are carried out with few resources, which is desirable so you can save your precious FPGA flip-flops and LUTs for other tasks.

Let's look at how the LFSR in Figure 6-4 could be implemented in Verilog and VHDL:

Verilog ❶
```
reg [2:0] r_LFSR;
wire      w_XNOR;
always @(posedge i_Clk)
```

```
      begin
    ❷ r_LFSR <= {r_LFSR[1:0], w_XNOR};
      end
  ❸ assign w_XNOR = r_LFSR[2] ^~ r_LFSR[1];
```

VHDL ❶ `signal r_LFSR : std_logic_vector(2 downto 0)`
```
        signal w_XNOR : std_logic;
        begin
          process (i_Clk) is
          begin
            if rising_edge(i_Clk) then
            ❷ r_LFSR <= r_LFSR(1 downto 0) & w_XNOR;
            end if;
          end process;
      ❸ w_XNOR <= r_LFSR(2) xnor r_LFSR(1);
```

First we declare a 3-bit-wide LFSR ❶. We perform the shift and incorporate the result of the XNOR operation through concatenation ❷. In Verilog we concatenate values by placing them in curly brackets, {}, separated by commas, while in VHDL we use a single ampersand (&). Together, the shifting and concatenation build up a single 3-bit-wide value, with w_XNOR in the least significant bit position. Finally, we assign the w_XNOR gate, based on the values of bits 2 and 1 in the register ❸. This is a continuous assignment, occurring outside the always or process block, and it will be implemented by a LUT in the FPGA.

NOTE *This example has shown a very simple 3-bit-wide LFSR, but an LFSR would normally have initialization and reset logic, which would help avoid and recover from any disallowed state. More thorough code, including reset logic and the ability to size the LFSR to any number of bits, is available in the book's GitHub repository.*

LFSRs are a simple and efficient way to perform several useful tasks. They also highlight one of the strengths of an FPGA, namely being able to quickly perform math operations with few resources. Consider that you could have hundreds of LFSRs running in parallel on a single FPGA without issue, and you can start to see how FPGAs excel at fast math operations running in parallel.

Project #5: Selectively Blinking an LED

Now that we've introduced some building blocks, let's start putting them together. The requirement for this project is to blink each of four LEDs on your development board on and off, but only one LED should be blinking at a time. You'll select which LED to blink using two switches. Table 6-6 shows how the LED selection is performed.

Table 6-6: LED Selection

i_Switch_2	i_Switch_1	LED to blink	Signal name
0	0	D1	o_LED_1
0	1	D2	o_LED_2
1	0	D3	o_LED_3
1	1	D4	o_LED_4

Looking at the table, we can see that when the two input switches are both 0 (not pressed), the D1 LED will blink. By pushing just switch 1 down (setting it to 1), we select the D2 LED to blink. When we push down just switch 2, D3 should blink, and finally, when we push down both buttons, D4 should blink. This sounds like a job for a demultiplexer! We'll have a single signal that toggles on and off, and we'll want to route it to one of four LEDs. But how can we generate the toggling signal?

The clock on a development board is quite fast. On the Go Board (discussed in Appendix A), for example, it's 25 MHz. If we fed that directly to an LED, then the LED would blink at 25 MHz. To the human eye, it would look like the LED was just on, since that's too fast for us to perceive. We need to generate a signal that toggles on its own, but at some much slower frequency than the clock: say, 2 to 4 Hz. That's fast enough that you'll be able to tell the LED is blinking quickly, but not too fast for the human eye to see. Remember, however, that FPGAs have no built-in concept of time, so we can't blink an LED by writing code like this:

```
r_LED <= 1;
wait for 0.20 seconds
r_LED <= 0;
wait for 0.20 seconds
```

As discussed in Chapter 5, an FPGA can determine how much time has passed by counting clock cycles. To wait for 0.20 seconds to pass, we would need to count one-fifth of the number of clock cycles that occur in a second. In the case of the Go Board, since there are 25,000,000 clock cycles per second (a 25 MHz clock), we would need to count to 25,000,000 / 5 = 5,000,000. Once the count hits this limit, we could reset it to zero and toggle the state of the LED.

But there's another way! Recall that one of the possible uses for an LFSR is to create a low-resource counter. Start an LFSR with a certain pattern, such as all zeros, and it will take $2^N - 1$ clock cycles for that pattern to recur, where N is the number of flip-flops that make up the LFSR. Create an LFSR with a high enough number of flip-flops in the shift register, and the rate at which it cycles through all its values will be slow enough to toggle the LED at a satisfying frequency. For example, a 22-bit LFSR will repeat its pattern every $2^{22} - 1 = 4{,}194{,}303$ clock cycles. With the Go Board's 25 MHz clock, that comes out to a little less than 0.20 seconds.

NOTE *If your board has a different clock frequency, you'll need to experiment with the number of bits in the LFSR. For the 100 MHz clock on the Alchitry Cu, for example (see Appendix A), try 24 bits: $2^{24} - 1 = 16{,}777{,}215$ cycles, or about 0.17 seconds.*

Each time the LFSR returns to all zeros, it will toggle a signal, and we'll use that signal to blink whichever LED is currently selected. All of this can be done using fewer FPGA resources than a traditional counter. Figure 6-5 shows a block diagram of how it will work.

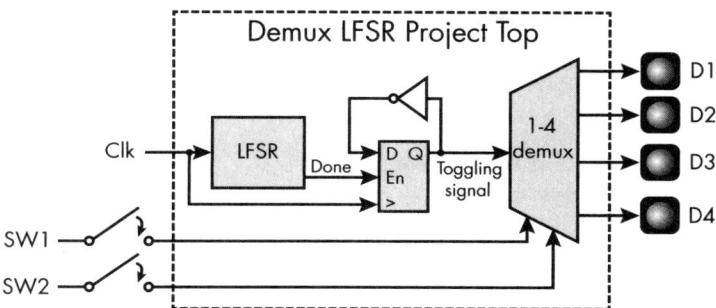

Figure 6-5: The Project #5 block diagram

This project will instantiate two modules: the LFSR and the 1-4 demux. Between the two modules, we'll have a flip-flop and a NOT gate (which will become a LUT). The input to the LFSR is the clock, and the output is a signal that goes high for one clock cycle when the LFSR has reached its limit and started at the beginning of its pattern again. We call this a *done pulse*. A *pulse* is a 1 (high) on a signal that lasts for one clock cycle, and this particular pulse signals when the LFSR is done with each cycle through its pattern loop.

We can't use the LFSR output directly to blink the LEDs, but we can use it to create a toggling signal. We do this by feeding the LFSR output signal into the enable input of a flip-flop. The flip-flop's output will be the inversion (using a NOT gate) of its input. This way, each time the LFSR cycles through its pattern, the done pulse will enable the flip-flop for one clock cycle and trigger a change on the flip-flop's output, either from a 0 to a 1 or from a 1 to a 0. The net result is a signal with a 50 percent duty cycle and a frequency of about 3 Hz, perfect for toggling an LED at a rate the human eye can see. This toggling signal is the input to the demux module. The 1-4 demux selects which LED to pass the toggling signal to by reading the values on the two switches (SW1 and SW2). Only one LED will be blinking at a time, while the LEDs not selected by the switches will be off.

Writing the Code

Let's look at the Verilog and VHDL for this project, starting with the top-level code:

Verilog

```verilog
module Demux_LFSR_Project_Top
 (input i_Clk,
  input i_Switch_1,
  input i_Switch_2,
  output o_LED_1,
  output o_LED_2,
  output o_LED_3,
  output o_LED_4);

  reg r_LFSR_Toggle = 1'b0;
  wire w_LFSR_Done;

❶ LFSR_22 LFSR_Inst
  (.i_Clk(i_Clk),
❷ .o_LFSR_Data(), // unconnected
❸ .o_LFSR_Done(w_LFSR_Done));
  always @(posedge i_Clk)
  begin
❹ if (w_LFSR_Done)
     r_LFSR_Toggle <= !r_LFSR_Toggle;
  end

❺ Demux_1_To_4 Demux_Inst
   (.i_Data(r_LFSR_Toggle),
    .i_Sel0(i_Switch_1),
    .i_Sel1(i_Switch_2),
    .o_Data0(o_LED_1),
    .o_Data1(o_LED_2),
    .o_Data2(o_LED_3),
    .o_Data3(o_LED_4));

endmodule
```

VHDL

```vhdl
library ieee;
use ieee.std_logic_1164.all;

entity Demux_LFSR_Project_Top is
  port (
    i_Clk     : in  std_logic;
    i_Switch_1 : in  std_logic;
    i_Switch_2 : in  std_logic;
    o_LED_1    : out std_logic;
    o_LED_2    : out std_logic;
    o_LED_3    : out std_logic;
    o_LED_4    : out std_logic);
end entity Demux_LFSR_Project_Top;

architecture RTL of Demux_LFSR_Project_Top is
```

```
        signal r_LFSR_Toggle : std_logic := '0';
        signal w_LFSR_Done    : std_logic;

begin

❶ LFSR_22 : entity work.LFSR_22
      port map (
        i_Clk      => i_Clk,
❷     o_LFSR_Data => open, -- unconnected
❸     o_LFSR_Done => w_LFSR_Done);

      process (i_Clk) is
      begin
        if rising_edge(i_Clk) then
❹       if w_LFSR_Done  = '1' then
            r_LFSR_Toggle <= not r_LFSR_Toggle;
          end if;
        end if;
      end process;

❺ Demux_Inst : entity work.Demux_1_To_4
      port map (
        i_Data  => r_LFSR_Toggle,
        i_Sel0  => i_Switch_1,
        i_Sel1  => i_Switch_2,
        o_Data0 => o_LED_1,
        o_Data1 => o_LED_2,
        o_Data2 => o_LED_3,
        o_Data3 => o_LED_4);

end architecture RTL;
```

Our project has three top-level inputs—the clock and two switches—as well as four outputs for the four LEDs. After declaring these, we instantiate the LFSR module ❶. We'll look closely at the module next, but for now, notice its o_LFSR_Done output ❸, which we wire to w_LFSR_Done. This output will pulse with each repetition of the LFSR loop.

We don't actually need the LFSR to output the current value on its register for this project, but this may be important in other contexts, so the LFSR module has an o_LFSR_Data output for this purpose. One handy trick when instantiating a module with unused outputs is to keep those outputs unconnected, which we do here with o_LFSR_Data ❷. In Verilog, we simply leave the parentheses after the output name empty, while in VHDL we use the open keyword. When this design is synthesized, the synthesis tool will prune any outputs that are unused, removing logic that doesn't go anywhere. This way, you can reuse modules without having to worry about devoting precious FPGA resources to unused features. The synthesis tools are smart enough to optimize your design and remove signals where they aren't needed.

In our top-level logic, we check if w_LFSR_Done is high, meaning the LFSR has output its done pulse ❹. If so, we invert the r_LFSR_Toggle signal. This is the signal that gets sent to the 1-4 demux, which we instantiate next ❺. The selection is performed by the two input switches, and the outputs of the demux are directly connected to the four output LEDs.

We've already seen the code for the 1-4 demux module, in "Implementing a Demultiplexer" on page 94. Let's look at the LFSR module now:

Verilog
```verilog
module LFSR_22 (
    input         i_Clk,
    output [21:0] o_LFSR_Data,
    output        o_LFSR_Done);

❶ reg [21:0] r_LFSR;
   wire       w_XNOR;

   always @(posedge i_Clk)
   begin
❷    r_LFSR <= {r_LFSR[20:0], w_XNOR};
   end

❸ assign w_XNOR = r_LFSR[21] ^~ r_LFSR[20];
❹ assign o_LFSR_Done = (r_LFSR == 22'd0);
❺ assign o_LFSR_Data = r_LFSR;

   endmodule
```

VHDL
```vhdl
library IEEE;
use IEEE.std_logic_1164.all;

entity LFSR_22 is
  port (
    i_Clk       : in std_logic;
    o_LFSR_Data : out std_logic_vector(21 downto 0);
    o_LFSR_Done : out std_logic);
end entity LFSR_22;

architecture RTL of LFSR_22 is

❶ signal r_LFSR : std_logic_vector(21 downto 0);
   signal w_XNOR : std_logic;

begin

   process (i_Clk) begin
     if rising_edge (i_Clk) then
❷      r_LFSR <= r_LFSR(20 downto 0) & w_XNOR;
     end if;
   end process;

❸ w_XNOR      <= r_LFSR(21) xnor r_LFSR(20);
❹ o_LFSR_Done <= '1' when (r_LFSR = "0000000000000000000000") else '0';
❺ o_LFSR_Data <= r_LFSR;

   end RTL;
```

This module is similar to the 3-bit LFSR that we looked at earlier in the chapter, but the LFSR register has been scaled up to be 22 bits wide ❶. (Modify the code if you need a different bit width based on your board's clock speed.) The module also has extra logic to generate the done pulse, as well as to output the LFSR data, which may be useful in other contexts.

We shift the LFSR register and concatenate the result with a new value for the rightmost bit ❷, just as we did in the 3-bit LFSR module. Then we XNOR the leftmost two bits in the register to get the new rightmost bit value ❸. We generate the done pulse on the o_LFSR_Done output when all of the flip-flops that make up the LFSR have zeros on their outputs ❹. Since this will be the case for exactly one clock cycle, this pulse will be one clock cycle wide. Otherwise, o_LFSR_Done will be low. Finally, we assign the contents of the LFSR register to the o_LFSR_Data output ❺. This way the module provides access to the LFSR data itself, but remember that in this case the o_LFSR_Data output won't be synthesized since we don't need the data for this particular application.

At this point, you can build and program the FPGA. When the project starts running, you should see one of the LEDs blinking, but you can select a different LED to blink by pushing either or both of the two switches.

Trying Another Way

This project has shown how simple building blocks like LFSRs and demuxes can be combined to build larger projects, and it has illustrated an interesting application for an LFSR. In the real world, however, you probably wouldn't use an LFSR to act as a counter like this, since it doesn't provide much flexibility. Let's say we want to change the count limit. With the LFSR implementation, we only have a few possible options to choose from, based on the number of bits in the LFSR. For blinking an LED, that was totally acceptable, as we didn't care exactly how fast the LED was blinking—anywhere between 2 and 4 Hz would be fine. But if we needed to count to a very specific value— say, 4,000,000 instead of 4,194,303—we'd be hard pressed to do this with the LFSR. The next lowest option would be to use a 21-bit LFSR instead of a 22-bit LFSR, which would only allow us to count to $2^{21} - 1 = 2,097,151$. For any value between 2,097,151 and 4,194,303, we're out of luck.

To provide more flexibility, I created another version of this project that uses a traditional counter. Figure 6-6 shows the block diagram of this alternate code.

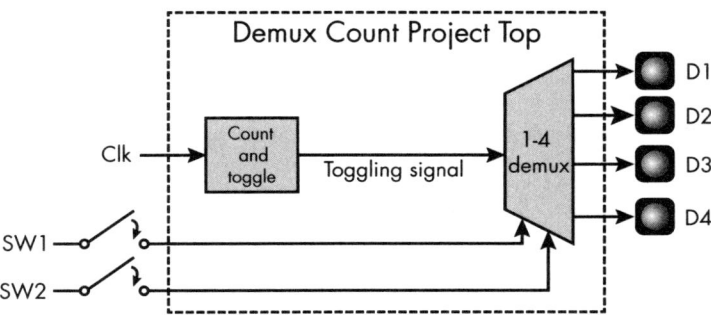

Figure 6-6: The revised Project #5 block diagram

Here, we've replaced the LFSR module with a module that simply counts up to some value and then toggles its output. This approach also allows us to eliminate the flip-flop and NOT gate between the project's two modules. Let's look at the code for the new Count_And_Toggle module:

Verilog

```verilog
module Count_And_Toggle #(COUNT_LIMIT = 10)
  (input i_Clk,
   input i_Enable,
   output reg o_Toggle);

❶ reg [$clog2(COUNT_LIMIT-1):0] r_Counter;

  always @(posedge i_Clk)
  begin
    if (i_Enable == 1'b1)
    begin
   ❷ if (r_Counter == COUNT_LIMIT - 1)
      begin
     ❸ o_Toggle  <= !o_Toggle;
     ❹ r_Counter <= 0;
      end
      else
     ❺ r_Counter <= r_Counter + 1;
    end
    else
      o_Toggle <= 1'b0;
  end

endmodule
```

VHDL

```vhdl
library ieee;
use ieee.std_logic_1164.all;
use ieee.numeric_std.all;

entity Count_And_Toggle is
  generic (COUNT_LIMIT : natural);
  port (
    i_Clk    : in std_logic;
    i_Enable : in std_logic;
    o_Toggle : out std_logic);
end Count_And_Toggle;

architecture RTL of Count_And_Toggle is

❶ signal r_Counter : natural range 0 to COUNT_LIMIT - 1;

begin

  process (i_Clk) is
  begin
    if rising_edge(i_Clk) then
      if i_Enable = '1' then
```

```
        ❷ if r_Counter = COUNT_LIMIT - 1 then
          ❸ o_Toggle  <= not o_Toggle;
          ❹ r_Counter <= 0;
            else
          ❺ r_Counter <= r_Counter + 1;
            end if;
          else
            o_Toggle <= '0';
          end if;
        end if;
      end process;

end RTL;
```

This code is much simpler to read and understand than the LFSR code. We declare a register that will act as a counter, using the COUNT_LIMIT parameter/generic to define its size ❶. If the module is enabled, we check if the counter has reached its limit ❷. If so, we invert the output signal ❸ and reset the counter ❹. If the counter isn't at its limit, then it simply increments by 1 ❺. With this code, we can set the counter to any arbitrary value and it will count to exactly that value.

NOTE   *In the VHDL code, we have an output signal o_Toggle on the right side of an assignment ❸, meaning we're accessing the output's value. This is valid in VHDL-2008 and later but will throw an error on older versions of VHDL. I recommend using VHDL-2008 in your designs, because of improvements like this.*

Now let's look at the changes to the top-level code that are needed to use this new Count_And_Toggle module instead of the LFSR:

Verilog
```
--snip--
    output o_LED_3,
    output o_LED_4);

    // Equivalent to 2^22 - 1, which is what the LFSR counted up to
    localparam COUNT_LIMIT = 4194303;

    wire w_Counter_Toggle;

❶ Count_And_Toggle #(.COUNT_LIMIT(COUNT_LIMIT)) Toggle_Counter
    (.i_Clk(i_Clk),
     .i_Enable(1'b1),
     .o_Toggle(w_Counter_Toggle));

    Demux_1_To_4 Demux_Inst
❷ (.i_Data(w_Counter_Toggle),
     .i_Sel0(i_Switch_1),
     .i_Sel1(i_Switch_2),
     .o_Data0(o_LED_1),
     .o_Data1(o_LED_2),
```

```
        .o_Data2(o_LED_3),
        .o_Data3(o_LED_4));

endmodule
```

VHDL
```
--snip--
architecture RTL of Demux_LFSR_Project_Top is

    -- Equivalent to 2^22 - 1, which is what the LFSR counted up to
    constant COUNT_LIMIT : integer := 4194303;

    signal w_Counter_Toggle : std_logic;

begin

❶ Toggle_Counter : entity work.Count_And_Toggle
    generic map (
      COUNT_LIMIT => COUNT_LIMIT)
    port map (
      i_Clk    => i_Clk,
      i_Enable => '1',
      o_Toggle => w_Counter_Toggle);

  Demux_Inst : entity work.Demux_1_To_4
    port map (
❷ i_Data  => w_Counter_Toggle,
      i_Sel0  => i_Switch_1,
      i_Sel1  => i_Switch_2,
      o_Data0 => o_LED_1,
      o_Data1 => o_LED_2,
      o_Data2 => o_LED_3,
      o_Data3 => o_LED_4);

end architecture RTL;
```

I've snipped the parts that are the same. The LFSR has been removed and replaced with the Count_And_Toggle module ❶. Since that module generates a toggling signal, we no longer need the flip-flop between the two modules. Instead, we can feed w_Counter_Toggle, the output of the Count_And_Toggle module, directly into the demux ❷.

Comparing the Two Approaches

As you've seen, using a traditional counter is simpler and more flexible than using an LFSR. However, earlier I asserted that implementing an LFSR requires fewer resources than a traditional counter. Let's compare the resource utilization reports for the two approaches to this project to see how significant the resource savings are. First, here's the report for the LFSR version:

```
--snip--
Register bits not including I/Os:  23 (1%)

Mapping Summary:
Total  LUTs: 13 (1%)
```

And here's the report for the counter version:

```
--snip--
Register bits not including I/Os:  24 (1%)

Mapping Summary:
Total  LUTs: 36 (2%)
```

The LFSR approach has used 1 fewer flip-flop and 23 fewer LUTs than the counter, so the LFSR does indeed require fewer resources. However, it helps to put that into perspective. Modern FPGAs have thousands of LUTs. You really shouldn't have to count every single one. By going with the LFSR, we might save 1 percent (or less) of the total resources of our FPGA, but we lose readability and flexibility in the design. In general, I prefer to implement solutions that make sense and are simple, and in this case the LFSR isn't the simplest solution.

In addition to showing you how to blink an LED and create a sophisticated project by combining various basic building blocks, this project has illustrated that there's often a trade-off between simplicity and resources. You'll find that there are typically several ways to solve problems within an FPGA, and you'll have to determine which solution works best for you. It could be that the most resource-efficient solution isn't the simplest, but on the other hand, the simplest solution may not require significantly more resources. In many cases, you might iterate on a design with different approaches, testing each one out. This is always a good exercise; you'll become a stronger FPGA engineer when you explore multiple ways to write your code.

Random Access Memory

Random-access memory (RAM) allows you to store data within your FPGA and read it back later. This is an incredibly common requirement in an FPGA design. For example, you might want to store data received from a camera, a computer, or a microcontroller and retrieve it later for processing, or you may need to create a storage space for data before saving it to a microSD card. These are just a few examples of use cases for a RAM. The *random-access* part of the name means that you can access the data in any order. On one clock cycle, for example, you could read out the first location of memory, and then on the very next clock cycle you could read out the last location of memory.

A RAM is typically designed to be either *single-port* or *dual-port*. In a single-port RAM, there's just one interface into the memory, so in a single clock cycle you can either read from or write to the memory, but not both. A dual-port RAM allows you to read from and write to the memory in the same clock cycle. The latter is more versatile and used more often, so we'll focus on how to implement that on an FPGA. Figure 6-7 shows at a high level what we'll be creating. Note that this is just one possible implementation; the exact signal names can vary.

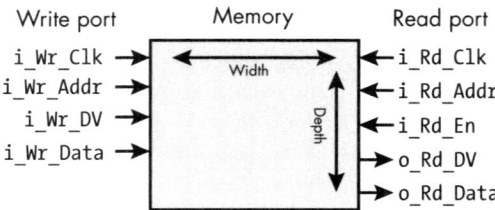

Write port   Memory   Read port

Figure 6-7: A dual-port RAM block diagram

In the middle of the figure, the memory itself is represented by the large rectangle. The size of the memory is defined by its width and depth. The depth determines the number of memory locations available, while the width determines how many bits can be stored at each location. For example, if the memory is 8 bits wide, then each location can store a byte of data. Multiplying the width by the depth tells you the total number of bits of memory available. For example, if we have an 8-bit-wide memory that's 16 locations deep, then there's a total of $8 \times 16 = 128$ bits of memory.

The memory has two ports, one for writing (on the left) and one for reading (on the right). Each port has its own clock signal, i_Wr_Clk and i_Rd_Clk. For our purposes, we'll tie both of these to the same clock, but note that it's possible for each port to operate according to its own independent clock. We'll discuss working with more than one clock, or *crossing clock domains*, in Chapter 7. For now, just know that this module is built with that feature in mind.

Each port has an address signal, i_Wr_Addr and i_Rd_Addr, which communicates the index into the memory where the writing or reading operation should take place. If you've programmed in C, this is like the index into an array. The indices typically range from 0 to (*depth* – 1), giving us a total of *depth* locations in the physical memory.

For writing the data, we need to set the write address correctly, put the data that we want to write on i_Wr_Data, and pulse the i_Wr_DV for a single clock cycle. DV here stands for *data valid*, which I commonly use to indicate that the data signal should be "looked at" by the module. If we want to keep writing to the memory, we can change the address and the data, and keep pulsing the data valid signal.

For reading the data, we drive i_Rd_En high, while setting the read address to the address we want to read from. The module that is performing the read can simply monitor the output o_Rd_DV to see when it goes high; this indicates that there is valid data on o_Rd_Data, which is the data that is read from the memory.

A RAM Implementation

Now that you understand at a high level how a RAM works, let's examine the code implementing the memory:

Verilog

```verilog
module RAM_2Port ❶ #(parameter WIDTH = 16, DEPTH = 256)
  (
  // Write signals
  input                     i_Wr_Clk,
  input [$clog2(DEPTH)-1:0] i_Wr_Addr,
  input                     i_Wr_DV,
  input [WIDTH-1:0]         i_Wr_Data,
  // Read signals
  input                     i_Rd_Clk,
  input [$clog2(DEPTH)-1:0] i_Rd_Addr,
  input                     i_Rd_En,
  output reg                o_Rd_DV,
  output reg [WIDTH-1:0]    o_Rd_Data
  );

❷ reg [WIDTH-1:0] r_Mem[DEPTH-1:0];

  always @ (posedge i_Wr_Clk)
  begin
❸ if (i_Wr_DV)
    begin
❹ r_Mem[i_Wr_Addr] <= i_Wr_Data;
    end
  end

  always @ (posedge i_Rd_Clk)
  begin
❺ o_Rd_Data <= r_Mem[i_Rd_Addr];
❻ o_Rd_DV   <= i_Rd_En;
  end

endmodule
```

VHDL

```vhdl
library ieee;
use ieee.std_logic_1164.all;
use ieee.numeric_std.all;

entity RAM_2Port is
❶ generic (
    WIDTH : integer := 16;
    DEPTH : integer := 256
    );
  port (
    -- Write signals
    i_Wr_Clk  : in std_logic;
    i_Wr_Addr : in std_logic_vector; -- sized at higher level
    i_Wr_DV   : in std_logic;
    i_Wr_Data : in std_logic_vector(WIDTH-1 downto 0);
```

```
      -- Read signals
      i_Rd_Clk  : in std_logic;
      i_Rd_Addr : in std_logic_vector; -- sized at higher level
      i_Rd_En   : in std_logic;
      o_Rd_DV   : out std_logic;
      o_Rd_Data : out std_logic_vector(WIDTH-1 downto 0)
      );
end RAM_2Port;

architecture RTL of RAM_2Port is

   type t_Mem is array (0 to DEPTH-1) of std_logic_vector(WIDTH-1 downto 0);
❷ signal r_Mem : t_Mem;

begin

   process (i_Wr_Clk)
   begin
     if rising_edge(i_Wr_Clk) then
   ❸ if i_Wr_DV = '1' then
       ❹ r_Mem(to_integer(unsigned(i_Wr_Addr))) <= i_Wr_Data;
       end if;
     end if;
   end process;

   process (i_Rd_Clk)
   begin
     if rising_edge(i_Rd_Clk) then
     ❺ o_Rd_Data <= r_Mem(to_integer(unsigned(i_Rd_Addr)));
     ❻ o_Rd_DV   <= i_Rd_En;
     end if;
   end process;

end RTL;
```

We've implemented the memory as a module called RAM_2Port. Notice that the module has two parameters (in Verilog) or generics (in VHDL): WIDTH and DEPTH ❶. This gives us the flexibility to create a RAM of any size we want, without having to modify the module code. If we need a memory that's 4 bits wide and 16 locations deep, this code can do that; if we need it to be 16 bits wide by 1,024 deep, this code can do that too. We only need to choose different WIDTH and DEPTH values when we instantiate the module.

Looking at the signal declarations to the module, we can see all the signals shown in Figure 6-7 that make up the write and read interfaces. The signals i_Wr_Addr and i_Rd_Addr will provide the indices of the write and read locations, respectively. These address signals are given a bit width large enough to represent any index to a memory containing DEPTH elements. For example, if you need to address into 128 memory locations (DEPTH = 128), then you're going to need 7 bits to accomplish that ($2^7 = 128$), so the address signals will be 7 bits wide. In Verilog, this sizing of the address works with the $clog2() trick described in Chapter 5. In VHDL we can leave the length of the vector undefined and set it in the higher-level module when this memory is instantiated. The instantiation itself must be of a fixed width, which will then specify the address signal width in this module. The last place we are using the parameters/generics to size our signals is for i_Wr_Data and o_Rd_Data. These carry the actual data being written or read, respectively, and are sized based on WIDTH to accommodate the full width of each location in memory.

We instantiate the memory itself as r_Mem ❷. It will be WIDTH wide and DEPTH long, for a total storage of WIDTH × DEPTH bits of memory. This instantiates a two-dimensional (2D) array in the code. In Verilog, we create it by setting a register of specific width, as we've done in the past, but with extra brackets on the end that specify the number of memory locations based on the DEPTH. In VHDL, we need to create a custom data type called t_Mem that defines the 2D array; then we can create the memory signal r_Mem, of type t_Mem.

Next, we give the write and read operations their own always or process blocks, triggered by the i_Wr_Clk and i_Rd_Clk clock signals, respectively. (Again, unless you need to cross clock domains, you can simply tie these signals to the same clock in the higher-level code that instantiates this module.) For write operations, we first check to see that the i_Wr_DV signal is high ❸. If it is, we take the data that's on i_Wr_Data and store it into memory at the location specified by i_Wr_Addr ❹. This looks a lot like updating a value in an array, because that's basically what we're doing.

For read operations, the o_Rd_Data output is updated with the value of the memory at the address given by i_Rd_Addr ❺. At the same time, the value on i_Rd_En is passed to o_Rd_DV ❻. The higher-level module will set i_Rd_En to high when it's actually trying to read data, and passing this signal to o_Rd_DV generates a data valid pulse telling the higher-level module that the data is safe to read. Notice, however, that i_Rd_En doesn't really control when data will be read within this module. In fact, the code to update o_Rd_Data ❺ will run on every single clock cycle, updating it with whatever is stored at the i_Rd_Addr memory location, whether we're explicitly trying to read data out of the memory or not. That's fine! It does no harm to read the memory on every clock cycle like this, even if we end up ignoring the data that's being read out.

To see the dual-port memory operating in a simulation, download the code from the repository and run the testbench for this module.

RAM on an FPGA

We've written the code for a dual-port RAM, but what FPGA component makes up the memory itself? The answer is, *it depends*. If the memory is small enough—for example, 4 locations wide by 8 deep—the storage elements will be individual flip-flops. However, if the memory gets large enough, the synthesis tools will instead decide to use a block RAM (BRAM). We'll discuss the block RAM in detail in Chapter 9. For now, just know that it's a large memory storage component that exists on the FPGA for this very purpose.

You wouldn't want to use flip-flops for large memories because you're limited with how many flip-flops are available for memory storage. You want to save those precious flip-flops to do the main work in your FPGA, not just store a single bit of data in a large memory. The synthesis tools are smart;

they know it's best to push a large memory instantiation to one or more block RAMs.

FIFO: First In, First Out

The *first in, first out (FIFO)* is another common FPGA building block for storing and retrieving data. The concept of a FIFO is quite simple: data comes in one entry at a time and gets read out in order from oldest to newest. Figure 6-8 shows a high-level representation of a FIFO.

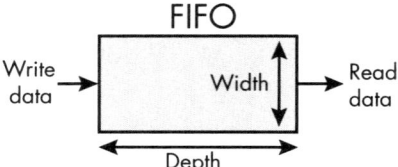

Figure 6-8: A high-level block diagram of a FIFO

As you can see, a FIFO has a write interface pushing data in on the left side and a read interface pulling data out on the right side. Compared to the dual-port RAM diagram in Figure 6-7, notice that I've flipped the width and depth here. This helps to visualize the key FIFO behavior: the first data that gets put in is the first data that gets pulled out. In this sense, data moves through a FIFO like cars through a tunnel. The first car into the tunnel is also the first car out. Other programming languages often have some sort of queue structure, which behaves the same way. With a FIFO in an FPGA, however, you're building a real queue out of real components!

FIFOs are used extensively in FPGA designs. Any time you need to buffer some data between a producer and a consumer, a FIFO acts as that buffer. For example, to write data to an off-chip memory storage element like a low-power double data rate (LPDDR) memory, you'd use many FIFOs to queue up the data, and then quickly burst it out of the FPGA into the LPDDR. Similarly, if you interface with a camera, you might store rows of pixel data into FIFOs for image manipulation like blurring or brightness enhancement. Finally, whenever you need to send data across clock domains, FIFOs are up to the task: one clock coordinates loading data into the FIFO, while the other clock coordinates reading it out.

A FIFO is full when there are no more memory locations available for new write data. A FIFO is empty when it has nothing in it. This leads to two critical rules that you must follow to ensure your FIFO behaves as expected:

1. Never write to a full FIFO.
2. Never read from an empty FIFO.

Writing to a full FIFO is bad because it can cause data loss: you'll end up overwriting data that was stored earlier. Reading from an empty FIFO is also bad, as you don't know what data you're going to get out of it. Breaking one of these two rules is one of the most common FPGA bugs that I've encountered in my career. It's also one of the harder bugs to find, because writing to a full FIFO or reading from an empty FIFO can cause strange behavior, like unexpected data and data loss. Often this corrupted data looks like a problem with the data, rather than a problem with the FIFO, so what's causing the bug is difficult to diagnose. Keep these rules in mind as we discuss the details of how a FIFO works.

Input and Output Signals

A FIFO is basically a version of a dual-port RAM with some extra signals added to create the FIFO behavior. Before we look at the code for the FIFO, let's consider what all those signals are. Figure 6-9 shows a more detailed block diagram of a FIFO.

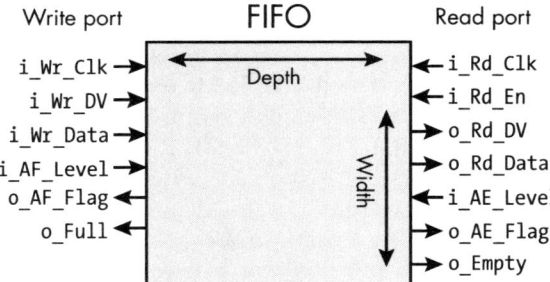

Figure 6-9: A detailed FIFO block diagram

Like the dual-port RAM, the FIFO has ports for writing and reading. Each port has its own dedicated clock. FIFOs are commonly used to cross clock domains, so i_Wr_Clk is different from i_Rd_Clk. However, the FIFO that we'll explore has a single clock for both the write and read ports, for simplicity and portability across FPGAs.

Next, on the write side, the i_Wr_DV (data valid) input signals when there's data to be written on i_Wr_Data and pushes that data into the FIFO. On the read side, the i_Rd_En and o_Rd_DV signals similarly communicate when we wish to read data, and the o_Rd_Data output retrieves the data itself. All of this is just like what we saw on the RAM. Unlike on the RAM, however, we no longer need to worry about keeping track of which address to write to or read from. The FIFO knows to simply cycle through the memory addresses in sequence, one after the other, when reading or writing. As such, there's no need for i_Wr_Addr and i_Rd_Addr input signals we had on the RAM. Instead, the remaining input and output signals help to track how much of the FIFO is used, while ensuring that we don't write to a full FIFO or read from an empty FIFO.

On the write side, the o_Full output goes high when all the locations within the FIFO have been written to. When the user sees o_Full go high, they must stop writing to the FIFO until some space frees up and o_Full goes low again. As you know by now, writing to a full FIFO is very bad and should be avoided.

The i_AF_Level and o_AF_Flag signals, also on the write side, aren't always included in FIFO modules, but they can be very helpful. AF is short for *almost full*, and these signals allow the user to set a watermark in the FIFO before it fills up completely. If the number of elements (sometimes called *words*) in the FIFO is such that i_AF_Level more words will not fit, then o_AF_Flag will be high. Otherwise, o_AF_Flag will be low. This feature is particularly useful in situations where data is written to the FIFO in batches. For example, say the write interface *must* write a minimum of four elements at a time in a burst, meaning once the interface starts writing it can't stop, even if the o_Full flag goes high mid-burst. To prevent data loss, we would want to set i_AF_Level to 4 and then check that o_AF_Flag is low before writing each burst of four elements. This ensures that there will be space in the FIFO for all four elements, before the write operation begins.

The read side has a similar set of FIFO-specific signals. o_Empty will be high when the FIFO has no data in it. To ensure that we never read from an empty FIFO, we should check the o_Empty flag before attempting to read data out to know if there's data available for reading.

The i_AE_Level and o_AE_Flag signals behave similarly to i_AF_Level and o_AF_Flag, guaranteeing that a burst of reads is possible without the FIFO becoming empty mid-burst (AE is short for *almost empty*). For example, say your FIFO is 1,024 bits deep and 1 byte wide, and you have an LPDDR interface that requires data to be written in 256-byte bursts. Again, since a burst can't be interrupted, you can't simply stop reading if the FIFO becomes empty partway through the read. To guarantee that at least 256 bytes will be available to be pulled from the FIFO before sending a burst of data to the LPDDR, set i_AE_Level to 256, and check that o_AE_Flag is low before reading data.

NOTE   *If you don't need almost full or almost empty behaviors for your application, you can just ignore the i_AF_Level, o_AF_Flag, i_AE_Level, and o_AE_Flag signals in your design.*

Figure 6-10 shows some examples summarizing what we've learned about FIFO signals.

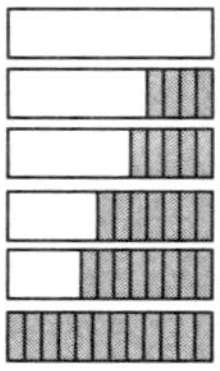

Count	o_Empty	o_AE_Flag	o_AF_Flag	o_Full
0	1	1	0	0
4	0	1	0	0
5	0	0	0	0
7	0	0	0	0
8	0	0	1	0
12	0	0	1	1

Figure 6-10: FIFO flag examples

The figure illustrates a FIFO with a depth of 12 words (the width isn't important). For this example, let's assume we've set i_AE_Level to 5 and i_AF_Level to 5. In the first row, we can see that if the FIFO has nothing in it, the count is zero and the o_Empty and o_AE_Flag signals are both set to 1. The o_AE_Flag is set when the count is less than i_AE_Level. Next, we see that there are four words written, the FIFO is no longer empty, but o_AE_Flag is still set. It's not until the fifth word is written that o_AE_Flag goes low. All flags are low from words five through seven, but when there are eight words in the FIFO, o_AF_Flag goes high (since i_AF_Level was set to 5, and 12 − 5 = 7). Said another way, we don't have enough space for 5 more words when the count is at 8. When the FIFO is full, we see that both o_AF_Flag and o_Full are high.

A FIFO Implementation

We'll now consider the Verilog and VHDL for implementing the FIFO illustrated in Figure 6-9. This code adds features around the RAM_2Port module discussed in "A RAM Implementation" on page 113 that convert the RAM into a FIFO. The complete code is available in the book's GitHub repository, along with the testbenches that exercise it. I'm not showing the module signals or the instantiation of the memory (the dual-port RAM from the previous section) here, so we can focus on the functional code that makes the FIFO a FIFO:

Verilog
```
--snip--
    always @(posedge i_Clk or negedge i_Rst_L)
    begin
❶ if (~i_Rst_L)
      begin
        r_Wr_Addr <= 0;
        r_Rd_Addr <= 0;
        r_Count   <= 0;
      end
      else
      begin
❷ if (i_Wr_DV)
        begin
          if (r_Wr_Addr == DEPTH-1)
            r_Wr_Addr <= 0;
          else
            r_Wr_Addr <= r_Wr_Addr + 1;
        end

❸ if (i_Rd_En)
        begin
          if (r_Rd_Addr == DEPTH-1)
            r_Rd_Addr <= 0;
          else
            r_Rd_Addr <= r_Rd_Addr + 1;
        end

❹ if (i_Rd_En & ~i_Wr_DV)
        begin
```

```
          if (r_Count != 0)
          begin
            r_Count <= r_Count - 1;
          end
        end

   ❺ else if (i_Wr_DV & ~i_Rd_En)
        begin
          if (r_Count != DEPTH)
          begin
            r_Count <= r_Count + 1;
          end
        end

        if (i_Rd_En)
        begin
          o_Rd_Data <= w_Rd_Data;
        end

      end // else: !if(~i_Rst_L)
    end // always @ (posedge i_Clk or negedge i_Rst_L)

  ❻ assign o_Full  = (r_Count == DEPTH) ||
                     (r_Count == DEPTH-1 && i_Wr_DV && !i_Rd_En);

    assign o_Empty = (r_Count == 0);

    assign o_AF_Flag = (r_Count > DEPTH - i_AF_Level);
    assign o_AE_Flag = (r_Count < i_AE_Level);

  --snip--
```

```
VHDL  --snip--
        process (i_Clk, i_Rst_L) is
        begin
    ❶ if not i_Rst_L then
          r_Wr_Addr <= 0;
          r_Rd_Addr <= 0;
          r_Count   <= 0;
        elsif rising_edge(i_Clk) then

      ❷ if i_Wr_DV then
            if r_Wr_Addr = DEPTH-1 then
              r_Wr_Addr <= 0;
            else
              r_Wr_Addr <= r_Wr_Addr + 1;
            end if;
          end if;

      ❸ if i_Rd_En then
            if r_Rd_Addr = DEPTH-1 then
              r_Rd_Addr <= 0;
            else
```

```
              r_Rd_Addr <= r_Rd_Addr + 1;
          end if;
        end if;

❹ if i_Rd_En = '1' and i_Wr_DV = '0' then
      if (r_Count /= 0) then
        r_Count <= r_Count - 1;
      end if;

❺ elsif i_Wr_DV = '1' and i_Rd_En = '0' then
      if r_Count /= DEPTH then
        r_Count <= r_Count + 1;
      end if;
    end if;

    if i_Rd_En = '1' then
      o_Rd_Data <= w_Rd_Data;
    end if;

  end if;
end process;

❻ o_Full <= '1' when ((r_Count = DEPTH) or
                      (r_Count = DEPTH-1 and i_Wr_DV = '1' and i_Rd_En = '0'))
                      else '0';
  o_Empty <= '1' when (r_Count = 0) else '0';

  o_AF_Flag <= '1' when (r_Count > DEPTH - i_AF_Level) else '0';
  o_AE_Flag <= '1' when (r_Count < i_AE_Level) else '0';
--snip--
```

The bulk of this code is the main always block (in Verilog) or process
block (in VHDL), which handles memory addressing, counting the number of elements in the FIFO, and read and write operations. Notice that
this block has a reset signal, i_Rst_L, in the sensitivity list, in addition to a
clock signal. If the reset signal goes low, then we're in a reset state and we
reset the signals that control the read address, the write address, and the
FIFO count ❶. The _L at the end of the reset signal name is a clue that it's
active-low.

NOTE *As I mentioned earlier, FIFOs are useful for crossing clock domains, but this particular implementation of a FIFO cannot do this. It only has one clock, the i_Clk signal.
Crossing clock domains is an advanced feature that we aren't prepared to implement
at this stage in the book.*

Next, we create the logic for the write address ❷ and read address ❸.
For these, we simply increment the address each time we do a write or a
read. When we reach the last address in the FIFO, which is DEPTH-1, we start
over again at 0. Thanks to this system, elements are written to memory
sequentially and they're read from memory in the same order they were
written, ensuring adherence to the first-in, first-out scheme.

To keep track of the number of elements in the FIFO, first we check for the condition where we're doing a read but not a write ❹. In this case, the total number of elements in the FIFO goes down by 1. Next we check if we're doing a write but not a read ❺, in which case the total number of elements in the FIFO increases by 1. It's also possible that we could be writing *and* reading at the same time, but notice that the code doesn't explicitly handle this case. That's intentional; in this situation, the count will remain the same. We could make this explicit by writing r_Count <= r_Count;, but this isn't necessary. By default, the count variable retains its value.

We also perform several signal assignments outside of the always or process block ❻. Recall that this will generate combinational (as opposed to sequential) logic. First we assign the o_Full flag, which will be high when r_Count is equal to DEPTH, or when r_Count is equal to DEPTH-1 *and* there's a write *and* there's not a read. This second case lets the full flag "anticipate" the write and tell the higher-level module to stop writing, since the FIFO is about to be full.

Next we have the o_Empty assignment, which is a bit simpler. When the count is zero, the FIFO is empty; otherwise, it's not empty. After that we assign the almost full (o_AF_Flag) and almost empty (o_AE_Flag) flags. For these, we need to compare the count of the FIFO to the thresholds determined by i_AF_Level and i_AE_Level, respectively. This is the first time we've seen the < and > comparison operators being used in Verilog and VHDL. These are perfectly valid to include in combinational signal assignments.

By monitoring these status flags from a higher-level module, you'll be able to precisely control when data can move into and out of the FIFO.

Summary

In this chapter, you learned about several common building blocks in FPGA designs, including multiplexers and demultiplexers, shift registers, RAM, and FIFOs. You saw how these components work and learned how to implement them with Verilog and VHDL. With these foundational pieces of code, you can start to see how very large FPGA designs can be composed of many smaller modules structured together.

7

SYNTHESIS, PLACE AND ROUTE, AND CROSSING CLOCK DOMAINS

In Chapter 2, I provided an overview of the FPGA build process to get you comfortable running the tools needed to work on this book's projects. We'll now take a closer look at the build process, to give you a deeper understanding of what exactly is going on when you click the Build FPGA button. Once you have a firm knowledge of what your FPGA tools are doing, you'll be able to avoid many common mistakes and write highly reliable code.

As you learned back in Chapter 2, after you've written your Verilog or VHDL code the FPGA design goes through three stages: synthesis, place and route, and programming. If any of these processes fails, the FPGA build won't be successful. In this chapter, we'll focus on the first two stages. We'll talk in detail about synthesis, and break down the differences between synthesizable and non-synthesizable code. After that, we'll revisit the place and route process and explore one common issue that arises during this stage: timing errors. You'll learn what causes these errors and how to fix them. Finally, we'll look in detail at a situation where you're particularly likely to encounter timing issues: when signals cross between parts of your FPGA design running at different clock frequencies. You'll learn how to safely cross clock domains in your FPGA.

Synthesis

Synthesis is the process of breaking down your Verilog or VHDL code and converting it to simple components (LUTs, flip-flops, block RAMs, and so on) that exist on your specific FPGA. In this sense, an FPGA synthesis tool is similar to a compiler, which takes code in a language like C and breaks it down into very simple instructions that your CPU can understand.

For the process to work correctly, the synthesis tool needs to know exactly what type of FPGA you're using so it knows what resources are available. Then, since these resources are finite, it becomes the synthesis tool's job to figure out how to use them as efficiently as possible. This is called *logic optimization* (or *logic minimization*), and it's a major part of the synthesis process. As I mentioned in Chapter 3, there's no reason for you to ever perform logic optimization manually; you can simply leave it to the synthesis tool. That's not to say, however, that writing code that uses your available resources intelligently isn't important. Having a good understanding of what your code will synthesize into is critical to becoming a strong FPGA designer.

A key output of the synthesis process is your *utilization report*, which tells you how many LUTs, flip-flops, block RAMs, and other resources you're using in your design. We've examined excerpts from utilization reports in past chapters; I recommend always reading through this report to make sure that your expectations match reality with regard to the resources being used.

Notes, Warnings, and Errors

The synthesis process often generates a very large number of notes and warnings, even when it runs successfully. When unsuccessful, the process generates errors as well. The notes are mostly informational, telling you how the tool is interpreting your code. Warnings are worth looking at to make sure that you're not making mistakes. However, in large designs there might be hundreds of warnings, so they can become overwhelming. Some tools allow you to hide warnings once you're comfortable with them. This is a useful feature that allows you to focus on the real problems.

One particular warning worth noting is the *inferred latch* warning. As you learned back in Chapter 4, latches are bad. They're often created accidentally, and the tools can have trouble analyzing them in the context of timing in your FPGA design. If you create a latch, you'll be notified of it during the synthesis process. You'll get a warning like this:

```
[Synth 8-327] inferring latch for variable 'o_test' [test_program.vhd:19]
```

Don't ignore this warning. Unless you're sure that you really want that latch in your design, you should try to remove it. I've been doing FPGA design for many years and I've never needed to use a latch, so you should have a very good reason if you're planning to keep it.

If something goes wrong during synthesis, you'll get an error rather than a warning. The two most common errors you'll encounter are syntax errors and utilization errors; we'll look at those next.

Syntax Errors

When you start the synthesis process, the first thing the tool will do is check your Verilog or VHDL code for syntax errors. These are by far the most common errors you'll encounter. There are literally hundreds of kinds of syntax errors that might be lurking in your code; perhaps you forgot to define a signal, mistyped a keyword, or left out a semicolon, for example. In the last case, you might see an error message like this one in Verilog:

```
** Error: (vlog-13069) design.v(5): near "endmodule": syntax error,
unexpected endmodule, expecting ';' or ','.
```

or this one in VHDL:

```
** Error: design.vhd(5): near "end": (vcom-1576) expecting ';'.
```

The synthesis tool will tell you on which line of which file it encountered the error. In the preceding error messages, for example, design.v(5) in Verilog or design.vhd(5) in VHDL is telling you to check line 5 of the file called *design*. You can use this information to edit your code to pass the syntax check.

Occasionally, you'll get an overwhelming number of syntax errors. The best thing you can do in this case is find the first error and fix that one. Often, a cascade of errors can stem from the first one. This is a good rule for engineering in general: fix the first problem first. Once you've resolved that first syntax error, rerun the synthesis process. If you're still getting errors, again find the first one, fix it, and rerun synthesis. This process is iterative, and it often takes a few cycles for all the syntax errors to be resolved and the synthesis process to complete successfully.

Utilization Errors

Once your code passes the syntax check, the next most common error you'll encounter during synthesis is a *utilization error*, where your design requires more components than the FPGA has available. For example, if your FPGA has 1,000 flip-flops but your code is requesting 2,000, you'll get a utilization error. The design simply won't fit on your FPGA, so you'll need to think of ways to shrink your code to instantiate fewer flip-flops. A good rule of thumb is to aim to utilize no more than 80 percent of the available

LUTs or flip-flops. This will make it easier for the place and route process to get your design to meet timing (more on this later in the chapter), as well as giving you more flexibility to modify your design or add new features in the future.

If you can't get your code to fit in your chosen FPGA, you have a few options:

1. Switch to a larger FPGA.
2. Identify the most resource-intensive modules and rewrite them.
3. Remove functionality.

Switching to a larger FPGA might be a big deal, but it isn't always. Many FPGA vendors offer higher-resource FPGAs in the same physical package as lower-resource parts. The higher-resource ones often cost a bit more, so you'll pay a few extra dollars for those additional resources, but the new FPGA won't take up any extra space on your circuit board. When you're selecting an FPGA for a project, it's a good idea to pick an FPGA family and package where you have the option to move up in resources, just in case you end up needing more resources than you expect.

If you can't switch to a different FPGA, the next step is to analyze your code to see if it uses more resources than necessary. This isn't a matter of low-level logic minimization to shave off a LUT here and a flip-flop there— the tools do that for you. Rather, there are ways in which you might inadvertently write code that uses dramatically more resources than you expect. As an example, I once traced a high utilization error to one single line of code that was dividing two numbers. As you'll learn in Chapter 10, division is often a very resource-intensive operation in an FPGA. I was able to change the division operation into a memory operation by creating a table of possible inputs and mapping each input to an output. This used a block RAM, but it freed up the LUTs and flip-flops used for the division and allowed the FPGA to pass the synthesis process. Rewriting code with a focus on lower resource utilization is a skill you'll sharpen as you gain more FPGA experience. You can dig into the utilization reports of each module to figure out which are using the most resources, and then examine those individually.

Another approach for reducing resource utilization is to have different inputs share the same FPGA resource. FPGAs often perform the same operation on multiple input channels. Instead of having dedicated FPGA resources for each channel, you can use a single implementation, with each channel taking a turn sharing the hardware. For example, say you have 100 channels that all need a cosine operation performed on them once a second. You could have channel 1 perform the cosine operation in the first 10 ms, then allow channel 2 to perform that same cosine operation in the next 10 ms, then channel 3, and so on. This way, the hardware used to perform the cosine operation can be shared between all the channels and only needs to be instantiated once, rather than being instantiated 100 times, once for each channel.

This keeps the overall resource utilization much lower, but it only works if you have the time available to share a resource. If your timelines are too tight this approach might not work. Additionally, it does add some complexity, because now you need to build a component that will negotiate the sharing. We refer to the process of sharing a resource as *arbitration*, and the component that performs sharing this is often referred to as an *arbiter*. Arbiters can be built to share off-FPGA resources as well. For example, we might have several modules that write data to a MicroSD card. An arbiter could be designed to allow those modules to share the single MicroSD card and prevent two modules from trying to write data at the same time, which would cause data loss or corruption.

If you've written very efficient code and you still can't make it fit on your FPGA, the only option left is to remove functionality. Maybe there's a microcontroller on the same board that can do some of the things that the FPGA was supposed to do. Or maybe you just need to tell your team that it won't work. FPGAs simply have a limit to the amount of stuff they can fit.

Non-synthesizable Code

Many keywords in the Verilog and VHDL languages can't be translated into FPGA components by a synthesis tool; they aren't synthesizable. If you include any of these keywords in your project code, the synthesis tool will simply ignore them. It might generate a warning or a note, but not an error. The tool will move forward with the synthesis process, omitting the non-synthesizable parts from the final design—and potentially leading to problems if you were counting on the functionality of the non-synthesizable code.

It might seem strange that non-synthesizable keywords exist in Verilog and VHDL, but they're useful for simulation and testbenches. As we discussed in Chapter 5, testing your code is critical, and the languages provide keywords to assist you with this. In fact, you can include non-synthesizable elements in your project code for simulation purposes and leave them in when you run the code through synthesis, since the tool will just ignore them. To be safe, you can explicitly tell the tool not to bother trying to synthesize these parts of your code by preceding them with `synthesis translate_off` and succeeding them with `synthesis translate_on`. This technique works for both Verilog and VHDL. For example, if you're designing a FIFO, you might want to assert that you're never writing to a full FIFO or reading from an empty FIFO when running simulations of the code. The `synthesis translate_off` and `synthesis translate_on` directives let you bake those assertions into the actual design code, without having to worry about maintaining separate code for simulation and synthesis.

Some of the most common areas where non-synthesizable code arises include keeping track of time, printing text, working with files, and looping. We'll consider those now.

Keeping Track of Time

As you know, there's no inherent way to measure the passage of time in an FPGA. Instead, we rely on counting clock cycles. Still, there are parts of both VHDL and Verilog that refer to time: for example, $time in Verilog or now in VHDL will provide a current timestamp, while a statement like #100 in Verilog or wait for 100 ns; in VHDL will create a short delay. These features can be useful for running simulations—for example, to trigger input signals at precise time intervals—but they aren't synthesizable.

Printing

One common way to get feedback during the testing process is to send text to the terminal. In C and Python, for example, you have functions like printf() and print() that will send text to a console to allow you to see what's going on. Similar functions exist in Verilog and VHDL. In Verilog, you can use $display() to send text to the terminal. In VHDL, it's a bit more complicated, and there are a few options. For example, you can use assert followed by report and severity note to send text to the screen, as shown here:

```
assert false report "Hello World" severity note;
```

These text outputs only work in simulation. They can't be synthesized, as the concept of a console or terminal doesn't exist on a physical FPGA.

Working with Files

In most cases, you can't synthesize Verilog or VHDL code that involves reading from or writing to a file. The FPGA has no concept of "files" or any operating system; you have to build all that stuff yourself if you really need it. Consider something like storing data from a temperature sensor. You might want to read data from the sensor every second and write those values to a file. This is possible to do in simulation with functions like $fopen() and $fwrite() in Verilog or file_open() and write() in VHDL, but in synthesis, forget about it.

One exception here is that some FPGAs allow you to use a text file to preload (initialize) a block RAM. The specifics of how different vendors accomplish this vary, so refer to the memory usage guide for your FPGA if this is something you ever need to do.

Looping

Loop statements *can* be synthesized, but they probably won't work the way you expect. You might be familiar with for loops from a software language like C or Python: they allow you to write concise code that repeats an operation a specific number of times one after another. In simulation, Verilog or VHDL for loops work this way. In synthesizable FPGA code, however, for loops work differently; they're used to condense replicated logic, providing a shorthand for writing several similar statements that are meant to be executed *at the same time, rather than one after another*. To demonstrate, consider this example code for a 4-bit shift register:

Verilog
```
always @(posedge i_Clk)
  begin
    r_Shift[1] <= r_Shift[0];
    r_Shift[2] <= r_Shift[1];
    r_Shift[3] <= r_Shift[2];
  end
```

VHDL
```
process (i_Clk)
begin
  if rising_edge(i_Clk) then
    r_Shift(1) <= r_Shift(0);
    r_Shift(2) <= r_Shift(1);
    r_Shift(3) <= r_Shift(2);
  end if;
end process;
```

Each clock cycle, this code shifts data through the r_Shift register. The value from bit 0 is shifted to bit 1, the value from bit 1 is shifted to bit 2, and so on. The assignment statements that accomplish this follow a completely predictable pattern: the value of r_Shift[i] is assigned to r_Shift[i+1]. Synthesizable Verilog and VHDL for loops provide a more compact way of writing predictable code like this. Using a for loop, we can rewrite the shift register code as follows:

Verilog
```
always @(posedge i_Clk)
  begin
  ❶ for(i=0; i<3; i=i+1)
    ❷ r_Shift[i+1] <= r_Shift[i];
  end
```

VHDL
```
process (i_Clk)
begin
  if rising_edge(i_Clk) then
  ❶ for i in 0 to 2 loop
    ❷ r_Shift(i+1) <= r_Shift(i);
    end loop;
  end if;
end process;
```

Here, we declare a for loop with incrementing variable i ❶. With each iteration, the statement assigning the value from bit i to bit i+1 ❷ is executed. For example, on the first iteration of the loop i is 0, so the line that gets executed is r_Shift[0+1] <= r_Shift[0]. The second time through i is 1, so we get r_Shift[1+1] <= r_Shift[1]. On the third and final iteration, we get r_Shift[2+1] <= r_Shift[2].

The important thing to realize here is that *this all happens in one clock cycle*. In effect, all iterations of the loop execute simultaneously, just as the three separate assignment statements in the version without the for loop will execute simultaneously. The two versions do exactly the same thing (and will synthesize to the exact same FPGA resources), except the for loop version is written in a more compact way.

A common mistake that beginners make is putting a for loop inside a clocked always or process block and expecting each iteration of the loop to take one clock cycle. Take the following snippet of C code, for example:

```
for (i=0; i<10; i++)
  data[i] = data[i] + 1;
```

Here we have an array, data, and we're incrementing every value inside the array by 1 using a for loop. (We're assuming here that data has 10 items.) If you try to do the same thing using a Verilog or VHDL for loop, expecting that it will take 10 clock cycles to run, you'll be very confused, since the loop will actually be executed in a single clock cycle. If you *do* want to run a sequence like this over a number of clock cycles, you can update the values inside an if statement that checks for an index value to exceed a certain threshold, like this:

Verilog
```
always @(posedge i_Clk)
  begin
❶ if (r_Index < 10)
    begin
❷   r_Data[r_Index] <= r_Data[r_Index] + 1;
❸   r_Index        <= r_Index + 1;
    end
  end
```

VHDL
```
process (i_Clk)
begin
  if rising_edge(i_Clk) then
❶ if r_Index < 10 then
❷   r_Data(r_Index) <= r_Data(r_Index) + 1;
❸   r_Index        <= r_Index + 1;
    end if;
  end if;
end process;
```

Here, we use an if statement to replicate the check that stops the for loop ❶. In this case, we want the operation to run 10 times, or until r_Index is no longer less than 10. (We're assuming that the index value starts at 0, although this isn't shown in the code.) Next, we increment a value in r_Data, using r_Index to access the correct item in the array ❷. Finally, we increment r_Index ❸, which will then be used on the next clock cycle to update the next value in the array. In total, this will take 10 clock cycles to execute. In general, when trying to write code that iterates like a conventional for loop, usually all you need to do is add a counter signal (like r_Index) and monitor it with an if statement, as you've seen here.

Until you're very confident in how FPGA for loops work, I recommend avoiding them in any synthesizable code.

Place and Route

Place and route is the process of taking your synthesized design and mapping it to physical locations on your specific FPGA. The place and route tool decides exactly which LUTs, flip-flops, and block RAMs (and other components we haven't talked about yet) in your FPGA will be used, and wires them all together. At the end of the process, you get a file that can be loaded onto the FPGA. As you've seen, actually programming the FPGA using this file is usually a separate step.

Place and route, as the name implies, is in fact two processes: the placement of the synthesized design into your FPGA, and then the routing of that design using physical wires to connect everything together. The routing process is often the most time-consuming step in the build process, especially for large designs. On a single computer, it can take several hours to route a complicated FPGA. This is one of the main reasons why simulations are critical. You may only get a few chances a day to test your design on an actual FPGA because the build process takes so long, so it's best to iron out as many problems as you can through simulation *before* starting this process.

Constraints

To run the place and route process, you need to constrain at least two aspects of your design: the pins and the clock(s). There are other elements that can be constrained as well—input/output delays, specific routing, and more—but these two are the most fundamental.

The pin constraints tell the place and route tool which signals in the Verilog or VHDL code are mapped to which physical pins on the FPGA. When you're working with your circuit board, you'll need to look at the PCB schematic to know which FPGA pins connect to switches, which pins connect to LEDs, and so on. This is an example of a situation in which having some knowledge of how to read schematics is helpful for an FPGA designer.

The clock constraints tell the tool about the clock frequency used to drive your FPGA (or frequencies, if you have multiple clock domains, as we'll discuss later in the chapter). Clock constraints are fundamental to the routing process, in particular, since there are physical limitations on how far a signal can travel and how much can be done to it within a single clock period. When the place and route process finishes, it will generate a timing report that takes the clock constraints into account. If everything is sure to work under the specified clock constraints, the design is said to meet timing, and the report will show this. If, however, the tool determines that the clock constraints may be too tight for what you've designed, the tool will display timing errors in your timing report. As you'll see next, timing errors are really bad!

Timing Errors

Timing errors occur when your design and clock constraints are asking for the FPGA components and wires to work at a faster pace than the place and route tool can guarantee they can handle. This means your FPGA might not work as desired. I say *might* because it's possible that it will work perfectly, despite the timing errors—there's no way to know for sure ahead of time. This is in part because an FPGA's performance is affected by its operating conditions; it can vary, for example, based on changes in voltage and temperature. It might sound odd that your FPGA will perform slightly differently at cold temperatures versus hot temperatures, but that's the reality.

It's the job of the place and route tool to stress your design and analyze how it will perform in all possible operating conditions, including worst-case scenarios. If the design can run at your specified clock frequency across all those conditions, the tool can guarantee that the FPGA will meet timing; otherwise, it will report timing errors. The tools won't stop you from programming your FPGA with a design that contains timing errors. Maybe they should, but they don't. Again, this is because it's uncertain whether or how the timing errors will manifest. The tools don't know if you'll be running your design on your desk at room temperature, or on a satellite in the vacuum of space. In either case, the design might work, or it might fail, or it might appear to work perfectly for five minutes before manifesting a small error. Timing errors can produce strange behavior.

I once was brought onto an FPGA design for a camera product that was riddled with timing errors that the previous designer hadn't bothered to fix. Instead, they had built the design, looked at the report and seen that it contained dozens of timing errors, and programmed the FPGA anyway. They then tested it at their desk to see if it worked. They ran it for a few minutes and didn't run into any problems, so they decided it was fine and integrated it into the product. Then the product started to fail in odd ways. Pixels would blink, or the scene would flicker, but only occasionally, so the user might ignore it. Even stranger, only some products would have issues, and the severity of the problem varied from unit to unit.

Once someone realized how bad the issue was, a serious effort was made to fix the timing errors and produce an FPGA design that would work 100 percent of the time. The FPGA tools had been trying to tell the original designer that there might be a problem. Not a *functional* problem—the code was theoretically OK as written—but there was a chance that it wouldn't work correctly with the given clock constraints under all operating conditions. The moral of this story is that when an FPGA acts in weird ways, it's very possible that you haven't looked closely at your timing report (or, as was the case with my prior coworker, ignored it completely!).

At their root, timing errors arise because FPGAs are subject to physical limitations. Up to this point, we've been working in an ideal world. We've imaged that all signals can travel instantly from their source to their destination, and that all flip-flops can change their output instantly when they see a rising clock edge. We've been assuming that if the code is correct, then everything will just work.

Welcome to reality! In the real world, nothing is truly instantaneous, and components behave in unpredictable ways if they're asked to work too quickly. Three physical limitations that contribute to FPGA timing errors are setup time, hold time, and propagation delay. Let's take a quick look at these, and then we'll explore how to fix timing errors.

Setup and Hold Time

Setup time is the amount of time for which the input to a flip-flop is required to be stable *before* a clock edge in order for the flip-flop to accurately register the input data to its output on that clock edge. *Hold time* is the amount of time for which the input must be stable *after* a clock edge in order for the flip-flop to reliably hold its current output value until the next clock edge. This is illustrated in Figure 7-1.

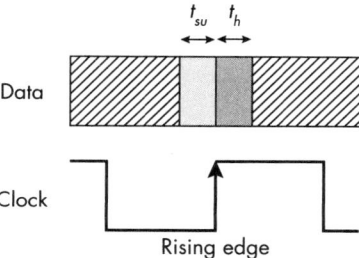

Figure 7-1: Setup (t_{su}) and hold (t_h) time

We expect the flip-flop to register some data at the rising clock edge shown in the middle of the figure. The time immediately before the rising edge is the setup time, labeled t_{su}. The time immediately after the rising edge is the hold time, labeled t_h. If the data input to the flip-flop changes outside the setup and hold window, then everything works fine. However, bad things can happen if your data input changes during the setup and hold window. Specifically, the flip-flop can become *metastable*, entering a state where its output is unstable: it could be a 1, it could be a 0, or it could even be somewhere in between. Figure 7-2 shows an example of a metastable event.

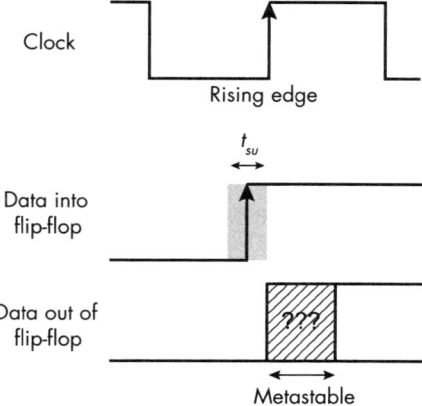

Figure 7-2: A metastable condition

Here we see a clock signal and the input and output signals of a flip-flop. The shaded area of the input signal labeled t_{su} represents the flip-flop's setup time, immediately before the rising clock edge. As you can see, the data input to the flip-flop transitions from low to high during the setup window. This causes the output to be metastable for some amount of time, after which it settles out to either a 0 or a 1.

To understand metastability, people often use the analogy of a ball balanced on top of a hill, as shown in Figure 7-3. The ball could roll down the hill either to the left or to the right, and there's no predicting which way it'll go. A random gust of wind could blow it one way or the other. If it rolls down to the left, that's state 0, and if it rolls to the right, that's state 1. When the output of a flip-flop is in a metastable state, it's a ball teetering on a hill, trying to find a more stable state to rest into.

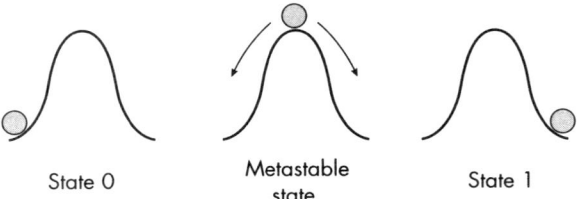

State 0   Metastable   State 1
                 state

Figure 7-3: A metastable state

In addition to not knowing which way the ball will roll, there's also no way to tell how long it will take for the ball to roll down the hill. It might fall quickly, or it might take a while. This is known as the *metastable resolution time*, or the time it takes for a metastable situation to become stable.

There's no way to know ahead of time which state the output will settle on. Sometimes it might be a 0, while other times when this situation occurs it might be a 1. Assuming the data input doesn't change again, the output will definitely be a 1 at the next rising clock edge, when the flip-flop again registers its input to its output. In the meantime, however, for the duration of this one clock cycle, there's no telling what the output will be, and this is not desired behavior for an FPGA.

If your design has timing errors, your FPGA tools are telling you that one or more flip-flops could have their setup and hold windows violated, which could put them in a metastable state. Metastability is probabilistic, however, so there's no guarantee that it will actually occur. There's a chance that your design will be completely fine despite the reported timing errors, but there's also a chance that the FPGA will exhibit strange and unpredictable behavior. In FPGA design we like things to be predictable, so even the remote possibility of metastability occurring is a problem.

Metastable conditions can occur when either the setup or the hold time is violated, but setup and hold time are physical properties of your FPGA and are beyond your control. You can't modify your design in a way that will change the setup or hold time. In order to resolve a timing error, you must focus your efforts on the other main physical limitation of FPGAs: propagation delay.

Propagation Delay

Propagation delay is the amount of time it takes for a signal to travel from a source to a destination. As mentioned previously, in the real world this is not instantaneous: it takes some time, albeit a very small amount, for voltage changes to propagate down a wire. A decent rule of thumb is that signals can travel along a wire at a rate of 1 foot per nanosecond. That may not sound like much of a delay, but consider that there are thousands of tiny wires running everywhere inside your FPGA. When you add up the physical length of the wires the total can be amazingly long, considering how small the chips are. This can lead to a significant propagation delay as signals travel from one flip-flop to another.

Additionally, every piece of logic that a signal goes through—for example, a LUT representing an AND gate—adds some extra time to the propagation delay, since these logic operations aren't perfectly instantaneous either. This concept is illustrated in Figure 7-4.

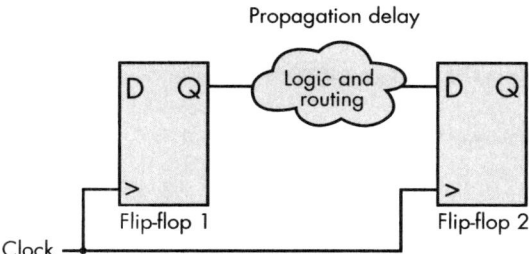

Figure 7-4: Propagation delay between two flip-flops

Here we have two flip-flops, with data traveling from the output of one flip-flop to the input of the other. The logic and routing in between might consist of wires and/or LUTs. This is where the propagation delay happens, and the more stuff there is in that cloud—for example, longer wires or more LUTs—the longer it will take for the output of flip-flop 1 to reach the input of flip-flop 2. If the propagation delay is too long, the design won't be able to meet timing at the requested clock constraint.

The issue here is that both flip-flops are driven by the same clock. If flip-flop 1 sees a change on its input and registers that change to its output at one rising clock edge, we'd expect flip-flop 2 to see that change and register it at the next rising clock edge. The signal only has a single clock period to propagate from flip-flop 1 to flip-flop 2. If the signal can safely arrive within that time, the design will work. But if the logic and routing in between the flip-flops create too long of a propagation delay, we'll get a timing error. There might be many thousands of flip-flops in the FPGA design, and it's the responsibility of the place and route tool to analyze every single path and show us the worst offenders from a timing perspective.

In fact, the signal has *less* than the length of a single clock period to propagate from flip-flop 1 to flip-flop 2, since we also need to take the setup time into account. The propagation delay may be less than the clock period, but as we've just seen, if the signal arrives at flip-flop 2 within its setup window, the output of flip-flop 2 will be uncertain. This leads to the following formula for calculating the clock period needed for a design to function properly:

$$t_{clk(min)} = t_{su} + t_p$$

Here, $t_{clk(min)}$ is the minimum clock period required for the design to work without timing errors, t_{su} is the setup time, and t_p is the worst propagation delay the design will experience between two flip-flops. As an example, say all of the flip-flops on the FPGA have a fixed setup time of 2 ns and our design will create a propagation delay of up to 10 ns (in the worst case) between two particular flip-flops. Our formula tells us that our clock needs to have a period of at least $2 + 10 = 12$ ns, which works out to a frequency of 83.3 MHz. We could easily run the design with a slower clock than that if we wanted, in which case the period would be even longer, but if we wanted to run the FPGA faster, say at 100 MHz, then the clock period would be too short and we would get timing errors.

How to Fix Timing Errors

As you've just seen, the clock period, setup time, and propagation delay are the main factors contributing to timing errors. Since the setup time is fixed, there are two basic ways to solve timing errors:

- Slow down the clock frequency.
- Reduce the propagation delay by breaking up the logic into stages.

Slowing down your clock frequency might seem like the most obvious choice. If you're able to run your FPGA slower, your timing will improve. However, it's unlikely that you'll be able to change your clock frequency freely; it's usually set in stone for some particular reason, such as if you're interfacing to a peripheral that needs to run at a specific frequency. Chances are you won't be able to slow the clock down just to relax timing.

Breaking up your logic into stages, also known as *pipelining*, is the more robust (and often the only) option. If you do less "stuff" between any two flip-flops, the propagation delay will decrease, and it will be easier for your design to meet timing. Figure 7-5 illustrates how this works.

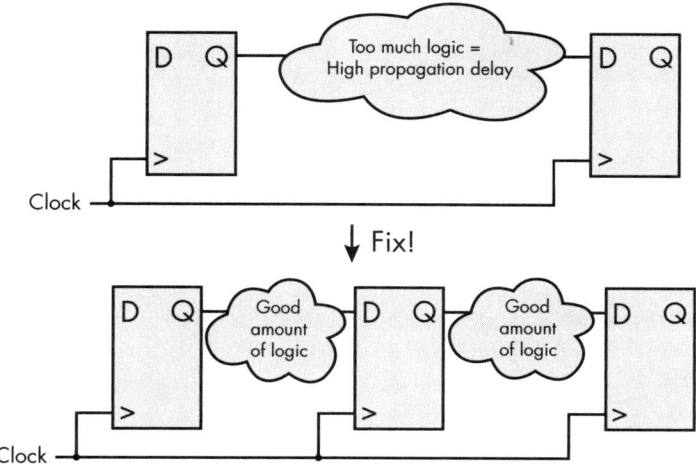

Figure 7-5: Reducing the propagation delay through pipelining

In the top half of the figure, we have a large amount of logic between two flip-flops—so much that the design has too long of a propagation delay and fails timing. The solution, shown in the bottom half of the figure, is to break up the logic into two stages, with another flip-flop added in between. This way, half of the logic can be done between flip-flops 1 and 2, and the other half between flip-flops 2 and 3. The propagation delay for each of these stages should be short enough that the stages can be accomplished in a single clock cycle, and overall the tools will have two clock cycles to do what we were originally trying to do in one clock cycle.

When you break up a single stage in your design into multiple stages like this, you're creating a *pipeline* of operations, with a flip-flop between each stage to synchronize the operations with the clock. A well-pipelined design will have a much better chance at meeting timing at high clock frequencies. To demonstrate, let's consider an example of some code that has poor timing, and then look at how to pipeline the logic to avoid timing errors. First, here's the problematic code:

Verilog
```verilog
module timing_error
  (input          i_Clk,
   input [7:0]    i_Data,
   output reg [15:0] o_Data);

  reg [7:0] r0_Data = 0;

  always @(posedge i_Clk)
  begin
    r0_Data <= i_Data;
❶ o_Data  <= ((r0_Data / 3) + 1) * 5;
  end
endmodule
```

```
VHDL  library ieee;
      use ieee.std_logic_1164.all;
      use ieee.numeric_std.all;

      entity timing_error is
        port (
          i_Clk  : in std_logic;
          i_Data : in unsigned(7 downto 0);
          o_Data : out unsigned(15 downto 0));
      end entity timing_error;

      architecture RTL of timing_error is
        signal r0_Data : unsigned(7 downto 0);
      begin

        process (i_Clk) is
        begin
          if rising_edge(i_Clk) then
            r0_Data <= i_Data;
  ❶        o_Data <= ((r0_Data / 3) + 1) * 5;
          end if;
        end process;

      end RTL;
```

I can't imagine why anyone would write code like this, but it will do for
demonstration purposes. The problem arises when we perform some mathematical operations—division, addition, and multiplication—on the value
of r0_Data ❶. All three of these operations are performed on the same line,
inside a synchronous always or process block, which means they must occur
within one clock cycle. To do all that math, the output of the 8-bit-wide
register r0_Data will pass through a bunch of LUTs, then into the inputs of
the flip-flops for o_Data, all within one clock cycle. This puts us firmly in the
upper half of Figure 7-5: the mathematical operations require a lot of logic
and will create a considerable propagation delay.

Let's see what happens when we run this code through place and route
with a 100 MHz clock constraint. Here's the resulting timing report:

```
  --snip--
  4.1::Critical Path Report for i_Clk
  **********************************
  Clock: i_Clk
❶ Frequency: 89.17 MHz | Target: 100.00 MHz
  ++++++++++++++++++++++++++++++++++++++++++++++++++++++++++++
❷ Path Begin : r0_Data_fast_5_LC_1_9_5/lcout
❸ Path End : o_DataZOZ_7_LC_5_12_5/in3
  Capture Clock : o_DataZOZ_7_LC_5_12_5/clk
  Setup Constraint : 10000p
❹ Path slack : -1215p
  --snip--
```

We can see that we tried to drive the clock to 100 MHz, but the place and route tool can only guarantee timing up to 89.17 MHz ❶. When the target frequency is higher than the maximum achievable frequency, we'll have timing errors. The timing report then tells us about the worst-offending paths in the design, albeit a little cryptically. First, the report identifies the beginning ❷ and end ❸ of each problematic path. Notice that r0_Data is in the signal name of Path Begin and o_Data is in the signal name of Path End, but there's a bunch of extra stuff there too. The tools add this additional information to identify the exact locations of the components in question within the FPGA. The downside is that the information isn't very human-readable, but since the core signal names have persisted, we can see that the path from r0_Data to o_Data is the failing path. Further, the report tells us exactly how much the path is failing by ❹. The *path slack* is the amount of wiggle room the path has available to meet timing, and the fact that it's negative is telling us we're too slow; we need an additional 1,215 picoseconds (ps), or 1.215 ns, to remove this timing error. That makes sense, since the difference in clock period between 89.17 MHz and 100 MHz is 1,215 ps.

Now that we've identified the failing path, we can pipeline the path's logic by breaking up the math operations with some flip-flops. Here's what that might look like:

Verilog
```
module timing_error
  (input           i_Clk,
   input [7:0]     i_Data,
   output reg [15:0] o_Data);

  reg [7:0] r0_Data, r1_Data, r2_Data = 0;

  always @(posedge i_Clk)
  begin
    r0_Data <= i_Data;
❶ r1_Data <= r0_Data / 3;
❷ r2_Data <= r1_Data + 1;
❸ o_Data  <= r2_Data * 5;
  end
endmodule
```

VHDL
```
library ieee;
use ieee.std_logic_1164.all;
use ieee.numeric_std.all;

entity timing_error is
  port (
    i_Clk  : in  std_logic;
    i_Data : in  unsigned(7 downto 0);
    o_Data : out unsigned(15 downto 0));
end entity timing_error;

architecture RTL of timing_error is
  signal r0_Data, r1_Data, r2_Data : unsigned(7 downto 0);
```

```
begin

  process (i_Clk) is
  begin
    if rising_edge(i_Clk) then
      r0_Data <= i_Data;
    ❶ r1_Data <= r0_Data / 3;
    ❷ r2_Data <= r1_Data + 1;
    ❸ o_Data  <= r2_Data * 5;
    end if;
  end process;

end RTL;
```

We've taken what used to be a single line of code and broken it up into
three lines. First, we perform just the division operation and write the result
to an intermediary signal, r1_Data ❶. Then we perform the addition operation
on r1_Data and assign the result to r2_Data ❷, and finally we perform the multiplication operation on r2_Data and assign the result to our original output,
o_Data ❸. We've introduced new signals to distribute the large math operation
that was occurring in a single clock cycle across multiple clock cycles. This
should reduce the propagation delay. Indeed, if we run the new, pipelined
design through place and route, we'll get the following timing report:

```
--snip--
4.1::Critical Path Report for i_Clk
**********************************
Clock: i_Clk
Frequency: 110.87 MHz | Target: 100.00 MHz
--snip--
```

Now we're meeting timing: the place and route tool can guarantee
performance up to a clock frequency of 110.87 MHz, when the target frequency was 100 MHz. As you've seen, fixing the timing errors involved a
bit of trade-off. We had to add flip-flops into our design to break up the
logic into stages, so our design now uses more FPGA resources than before.
Additionally, the math operation that was meant to take a single clock cycle
now takes three clock cycles. However, keep in mind that our module can
still take in new input values at a rate of 100 MHz and spit out calculated
output values at a rate of 100 MHz, as we intended with the original design;
it's only the *first* result to come out of the pipeline that will take two additional clock cycles, due to the added flip-flops.

Unexpected Timing Errors

You can fix most timing errors by pipelining your design to cut down on
propagation delay and avoid metastable conditions. However, place and
route tools can't anticipate every timing error. These tools aren't perfect;
they can only analyze your design based on the information they have. Even
if you don't see any errors in your timing report, there are two situations

where metastable conditions may still occur that the place and route tool can't reliably predict:

- When sampling a signal asynchronous to the FPGA clock
- When crossing clock domains

Sampling a signal asynchronous to the FPGA clock is very common when you have an external signal that serves as input to your FPGA. The input signal will go directly to the input of a flip-flop in your design, but it will be asynchronous, meaning it isn't coordinated by your main FPGA clock. For example, think about someone pushing a button. That button press could come at any time, so if it happens to occur during the setup time of the flip-flop the button is wired to, then that flip-flop will be in a metastable state. The place and route tool doesn't know anything about this potential issue, so it won't flag it as a timing error. But you, dear FPGA designer, can anticipate and fix the problem. The solution is to *double-flop* the input data by passing it through an extra flip-flop, as shown in Figure 7-6.

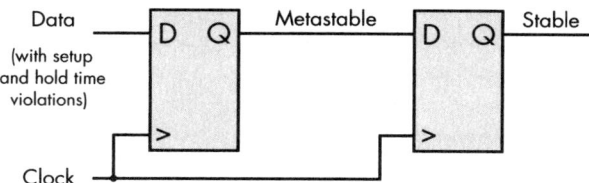

Figure 7-6: Double-flopping to fix metastability

In this figure, a signal that's asynchronous to the clock is being sampled by the first flip-flop. Since we can't guarantee the timing of this input relative to the clock, it might violate setup or hold times and create a metastable condition at the output of the first flip-flop. As discussed previously, though, the occurrence of a metastable condition is probabilistic. It's also quite rare. Even if the first flip-flop enters a metastable state, it's highly unlikely that the second flip-flop will enter one too. In fact, adding a second flip-flop in series with the first one reduces the likelihood of a metastable condition at the output to effectively zero. (Adding a third flip-flop in series would decrease the likelihood even more, but FPGA experts have concluded that two flip-flops in series is enough.) We can now use the stable signal internal to our design and be confident that we won't see strange behavior.

The other situation where you still may encounter a metastable state is when you cross clock domains in your FPGA. This is a big topic that warrants its own section.

Crossing Clock Domains

As I've mentioned, it's possible to have multiple clock domains inside a single FPGA, with different clocks driving different parts of the design. You might need a camera interface running at 25.725 MHz and an HDMI interface running at 148.5 MHz, for example. If you wanted to send the data from the camera out to the HDMI for visualization on a monitor, that data would have to cross clock domains, moving from the part of the design controlled by a 25.725 MHz clock to the part of the design controlled by a 148.5 MHz clock. However, there's no way to guarantee the alignment between these clock domains; they can drift apart and back together. Even if the clocks have a seemingly predictable relationship, like a 50 MHz clock and a 100 MHz clock, you can't be sure that the clocks started at the same time.

NOTE *The exception is if you're using an FPGA component called a* phase-locked loop (PLL), *which can generate unique clock frequencies and establish relationships between them. The PLL is discussed in detail in Chapter 9.*

The bottom line is that when you have clock domains that are asynchronous to each other, signals crossing between domains might produce metastable states within some flip-flops. In this section, we'll look at how to cross clock domains safely, from slower to faster and vice versa, and avoid metastability. We'll also discuss how to use a FIFO to send large amounts of data across clock domains.

Crossing from Slower to Faster

The simplest situation to handle is going from a slower clock domain to a faster clock domain. To avoid problems, all you need to do is double-flop the data when it enters the faster domain, as shown in Figure 7-7. This is the same approach we took to fix the external asynchronous signal, since it's fundamentally the same issue: the signal from the slower clock domain is asynchronous to the faster clock domain that it's entering.

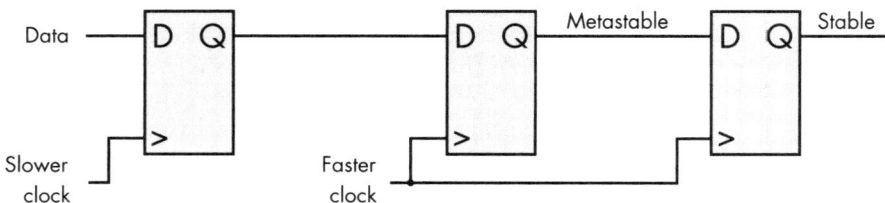

Figure 7-7: Crossing from a slower to a faster clock domain

In Figure 7-7, we have three flip-flops in series. The first is driven by a slower clock, while the others are driven by a faster clock. The slower clock is your *source clock domain*, and the faster clock is your *destination clock domain*. Since the clocks are asynchronous to each other, we can't guarantee that data coming from the slower clock domain won't violate the setup or hold times of the middle flip-flop (the first one in the faster clock domain) and trigger a metastable state. However, we know that the output of the second of these flip-flops will be stable, allowing the data to be used in the faster clock domain. Let's look at how this design could be implemented in code:

Verilog
```
always @(posedge i_Fast_Clk)
  begin
  ❶ r1_Data <= i_Slow_Data;
  ❷ r2_Data <= r1_Data;
  end
```

VHDL
```
process (i_Fast_Clk) is
begin
  if rising_edge(i_Fast_Clk) then
  ❶ r1_Data <= i_Slow_Data;
  ❷ r2_Data <= r1_Data;
  end if;
end process;
```

The code consists of an always or process block running off the positive edges of the faster clock. First, the signal i_Slow_Data (coming from the slower domain) enters flip-flop r1_Data ❶. The output of this flip-flop could be metastable if a change in i_Slow_Data violates its setup or hold time, but we resolve this metastable condition by double-flopping the data, passing it through a second flip-flop, r2_Data ❷. At this point, we have stable data that we can use in the faster clock domain without having to worry about metastable conditions.

One word of caution about writing code for an FPGA that uses two clock domains: be very careful to keep the code for the two domains separate. Keep all of your slower signals in one always or process block, clearly separated from the faster signals in a different always or process block (the exception is signals crossing between domains). In fact, I find it helpful to put code that runs in different clock domains in completely different files, just to be sure I'm not mixing and matching signals.

Crossing from Faster to Slower

Going from a faster clock domain to a slower one is more complicated than the other way around, since the data inside the faster clock domain could easily change before the slower clock domain even sees it. For example, consider a pulse that occurs for one clock cycle in a 100 MHz clock domain, that you're trying to detect in a 25 MHz clock domain. There's a good chance that you'll never see this pulse, as illustrated in Figure 7-8.

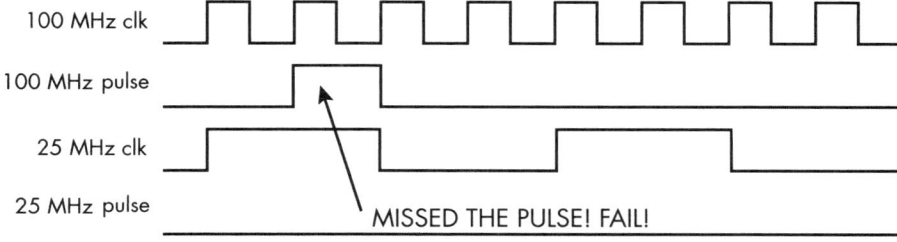

Figure 7-8: A failed crossing from a faster to a slower clock domain

This figure shows two cycles of the 25 MHz clock. After the clock's first rising edge but before its second rising edge, the 100 MHz pulse comes and goes, so fast that the 25 MHz clock never "sees" and registers it. This is because the pulse doesn't occur during the rising edge of the 25 MHz clock. Therefore, the pulse will go completely unnoticed in the 25 MHz clock domain. The solution to this problem is to *stretch out* any signals from the faster clock domain that are meant to enter the slower clock domain, until they're long enough to guarantee that they'll be noticed. Figure 7-9 shows how this works.

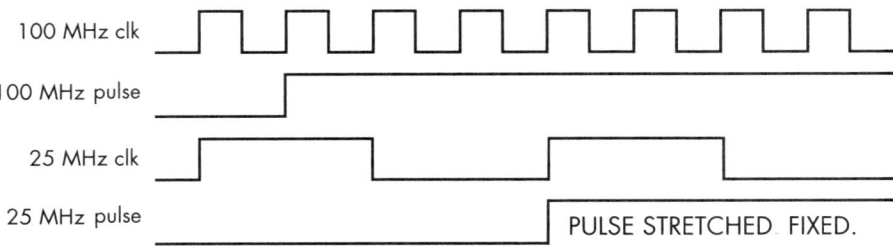

Figure 7-9: A successful crossing from a faster to a slower clock domain

Looking at the new waveform of the 100 MHz pulse, we can see that it's been stretched from a single cycle of the 100 MHz clock to multiple clock cycles, ensuring that a rising edge in the 25 MHz clock domain will see the pulse. As a general rule, pulses passing from a faster clock domain to a slower clock domain should be stretched to last at least two clock cycles in the slower domain. This way, even if the pulse violates the setup and hold time of the first clock cycle in the slower domain and triggers a metastable state, it will be stable at the second clock cycle. In our example, we should stretch out the 100 MHz pulse to at least eight 100 MHz clock cycles, the equivalent of two 25 MHz clock cycles. You can stretch the signals even longer if you like.

Using a FIFO

In the previous examples we looked at how to transmit a simple pulse across clock domains. But what if you want to send a lot of data between two clock domains, such as when sending camera data to an HDMI interface? In this case, the most common method is to use a FIFO. You write data to the FIFO according to one clock and read it out according to the other. Of course, the key requirement here is that the FIFO must support two different clock frequencies, whereas the FIFO we looked at in Chapter 6 only supported one clock.

To cross clock domains with a FIFO, you'll likely need to use a *primitive*, a dedicated FPGA component designed by the manufacturer specifically for your exact FPGA. For example, Intel will have prewritten FIFO primitives that run in different clock domains, but the Verilog or VHDL code to create them will differ from that of AMD's FIFO primitives. (We'll explore more examples of primitives in Chapter 9.)

When using FIFOs, always remember not to violate the two cardinal rules: don't read from an empty FIFO, and don't write to a full FIFO. Many

FIFO primitives provide a count of the number of words (elements) in the FIFO as an output, but I don't recommend relying on it. Instead, I suggest making heavy use of the *AF* and *AE* (almost full and almost empty) flags, introduced in "Input and Output Signals" on page 117. It's best to read from and write to the FIFO in bursts of a fixed size, especially when crossing clock domains, and to use that burst size to determine your AF and AE threshold levels. Set your AE level equal to your fixed burst size, and set your AF level equal to the FIFO depth minus your burst size. With this setup, you can guarantee that you'll never break the two key rules. The AF flag can throttle the write clock interface by ensuring that if the FIFO doesn't have enough room for another burst, writes will be disabled. Likewise, the AE flag can throttle the read clock interface by ensuring that if the FIFO doesn't have a complete burst in it, the read side will not try to pull data out of the FIFO.

Let's explore an example. Consider a case where we have some module writing data to a FIFO at 33 MHz. On the read side, we're dumping the data as quickly as possible to external memory. Let's say the read clock frequency is 110 MHz. In this situation, because the read clock is much faster than the write clock, the read side will be idle much of the time, even if the writes are happening at every clock cycle. To avoid reading from an empty FIFO, you can set the AE flag to some number that indicates to your read code that there's a burst of data ready to be read. If you set it to 50 words, for example, once the FIFO has 50 words inside it the AE flag will change from a 1 to a 0, which will trigger some logic on the read side to drain exactly 50 words from the FIFO.

This is often how crossing clock domains gets implemented. If you're using your AE/AF flags, you're doing it correctly. Try not to rely on the flags that tell you when the FIFO is completely full or empty, and definitely don't use the counter that some FIFOs support.

Addressing Timing Errors

When running a design with multiple clock domains through the place and route process, you'll need to include the frequencies of each clock in your clock constraints. The place and route tool will analyze data sent and received between these clock interfaces and report on any timing errors it observes. As you've seen, this is the tool telling you that you're likely to encounter situations where setup and hold times will be violated, which could trigger metastable conditions.

Let's assume you've handled the clock domain crossings well, using the double-flopping, data stretching, and FIFO methods discussed in the previous sections. You're aware of the possibility of the metastable states, and you've prepared for them. The tool doesn't know about the steps you've taken, though. It's unable to see that you're a smart FPGA designer and that you've got it all under control. To suppress the errors in your timing report, you'll need to create some unique *timing constraints* that relax the tools and tell them that you're a competent designer and that you understand that your design may be metastable. Exactly how to create these timing constraints is beyond the scope of this book, as each FPGA vendor has its own unique style.

You should always aim to have no timing errors in your design. However, as an FPGA designer, you'll inevitably experience situations where you cross clock domains. You need to clearly understand the common pitfalls that can occur in these situations. If the crossing is simple enough, you can just double-flop the data or perform data stretching. For many situations, you'll likely need to use a FIFO that supports two clocks, one for reading and one for writing. When structuring your code that handles crossing clock domains, be very careful not to mix and match signals from both clock domains.

Summary

In this chapter we've explored the FPGA build process in detail, looking closely at what happens to your FPGA code when it is synthesized and run through place and route. You've learned about different categories of non-synthesizable code, and in particular seen how for loops synthesize differently than you may expect. While examining the place and route process, you learned about some of the physical limitations of FPGAs and saw how they can lead to timing errors. Finally, you learned some strategies for fixing timing errors, including errors that can arise when crossing clock domains. With this knowledge, you'll be able to write your Verilog or VHDL code more confidently, and you'll be able to address common issues that arise during the build process.

8

THE STATE MACHINE

A *state machine* is a model for controlling a sequence of actions. In a state machine, a task is broken down into a series of stages, or *states.* The system flows through these states along prescribed routes, transitioning from one state to another based on inputs or other triggers. State machines are widely used to organize the operations in an FPGA, so understanding how they work is crucial to developing sophisticated FPGA designs.

Some common examples of state machines control behaviors in elevators, traffic lights, and vending machines. Each of these devices can only be in one unique state at any given time and can perform different actions as a result of inputs. In the case of an elevator, for example, the elevator car remains on its current floor until someone pushes a button to request a ride. The floor that the elevator is on is its state, and pushing a button is the input that triggers a change in that state. In the case of a traffic light, the possible states are red, yellow, and green, and the light changes based on some kind of input—perhaps a timer or a motion sensor. Certain transitions are possible, such as going from red to green or from yellow to red, while other transitions, such as yellow to green, are not.

Within an FPGA, you might have several state machines performing different independent tasks, all running simultaneously. You might have one state machine that initializes an LPDDR memory, another that receives data from an external sensor, and a third for communicating with an external microcontroller, for example. And since an FPGA is a parallel device, these state machines will all be running in parallel, with each state machine coordinating its own complicated series of actions.

In this chapter, you'll learn the basic concepts behind state machines and see how to design them with Verilog and VHDL. You'll learn strategies for keeping your state machines clear and concise, which reduces the likelihood of bugs in your designs. Finally, you'll gain hands-on experience with state machines by designing a Simon-style memory game for your development board.

States, Transitions, and Events

State machines revolve around three interrelated concepts: states, transitions, and events. A *state* describes the status of a system when it's waiting to execute a transition. Going back to the elevator example, if a button is never pressed, the elevator will simply remain in its current state; that is, waiting on its current floor. A state machine can only ever be in one state at a time (an elevator can't be on two floors at once), and there are only so many possible states that it can be in (we haven't yet figured out how to build a building with an infinite number of floors). For this reason, state machines are also called *finite state machines (FSMs)*.

A *transition* is the action of moving from one state to another. For an elevator, that would include opening and closing the doors, running a motor to raise or lower the car, and so on. Transitions are usually caused by *events*, which can include inputs like a button press or a timer expiring. Transitions between states can also occur without an external event, which would be an internal transition. The same event might trigger a different transition, depending on the current state of the state machine. For example, pushing the 5 button will make an elevator go down if it's on the tenth floor, or go up if it's on the first floor. The elevator's next state is influenced by both its current state and the input event.

Designing a state machine entails determining all the possible states, planning out the transitions between the states, and identifying the events that can trigger those transitions. The easiest way to do this is to draw a diagram. To demonstrate, let's explore an example of a simple state machine, one that controls a coin-operated turnstile like you might use to enter a subway station. Figure 8-1 shows a diagram of this state machine.

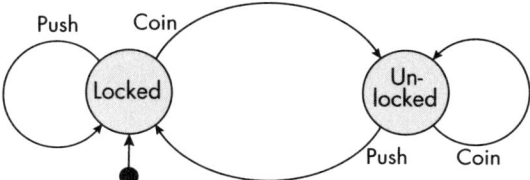

Figure 8-1: A state machine for a turnstile

In a state machine diagram, states are traditionally represented as labeled circles, transitions are represented as arrows between circles, and events are represented as text alongside the transitions they trigger. Our turnstile state machine has two possible states: Locked and Unlocked. The black dot beneath the Locked state indicates that Locked is the *initial state* of the machine. This is where the state machine will go when power is first applied, or if the user hits a reset button.

Let's consider what happens once we're in the Locked state. There are two possible events that can trigger a transition: pushing on the turnstile, or depositing a coin. A Push event causes a transition from Locked back to Locked, represented by the arrow on the left of the diagram. In this case, the state machine stays in the Locked state. It's not until a user deposits a coin (the Coin event) that we transition to the Unlocked state. At this point, if the user pushes the turnstile, it will let them through, then transition back to the Locked state for the next user. Finally, notice that if a user deposits a coin into a system that's already Unlocked, it transitions back into the Unlocked state.

It may seem trivial to define the behavior of a subway turnstile this way, but it's good practice for getting to know how state machines are organized and represented. For systems that have a large sequence of states, events, and transitions, explicitly documenting the state machine is critical for generating the desired behavior. Even something as simple as a calculator can require a surprisingly complex state machine.

Implementing a State Machine

Let's look at how to implement our basic subway turnstile state machine in an FPGA using Verilog or VHDL. We'll consider two common approaches: the first uses two always or process blocks, while the second uses just one. It's important to understand both of these approaches to implementing a state machine, since they're both widely used. However, as you'll see, there are reasons to prefer the latter approach.

Using Two always or process Blocks

Using two always or process blocks is a more traditional way of implementing a state machine. Historically, FPGA synthesis tools weren't very good. They could make mistakes when trying to synthesize state machines. The two-block approach was devised to get around these limitations. One always or process block controls the synchronous logic, using a register to keep track of the current state. The other always or process block controls the combinational logic; it looks for transition-triggering events and determines what the next state should be. Figure 8-2 illustrates this arrangement.

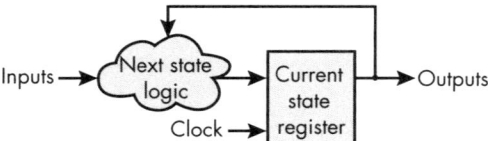

Figure 8-2: A block diagram of a state machine with two always or process blocks

Notice that the next state logic in this diagram doesn't have a clock as an input. It determines the next state immediately, based on the current state and any inputs (events). Only the current state register has a clock input, which it uses to register the output of the next state logic. In this way, it stores the current state of the machine.

Here's how to implement the turnstile state machine using this two-block approach:

Verilog

```
module Turnstile_Example
  (input i_Reset,
   input i_Clk,
   input i_Coin,
   input i_Push,
   output o_Locked);

❶ localparam LOCKED   = 1'b0;
   localparam UNLOCKED = 1'b1;

   reg r_Curr_State, r_Next_State;

   // Current state register
❷ always @(posedge i_Clk or posedge i_Reset)
   begin
     if (i_Reset)
     ❸ r_Curr_State <= LOCKED;
     else
     ❹ r_Curr_State <= r_Next_State;
   end

   // Next state determination
❺ always @(r_Curr_State or i_Coin or i_Push)
   begin
     r_Next_State <= r_Curr_State;

   ❻ case (r_Curr_State)

       LOCKED:
         if (i_Coin)
         ❼ r_Next_State <= UNLOCKED;
```

```
          UNLOCKED:
            if (i_Push)
              r_Next_State <= LOCKED;
         ❽
        endcase
      end
 ❾ assign o_Locked = (r_Curr_State == LOCKED);

 endmodule
```

VHDL
```vhdl
library ieee;
use ieee.std_logic_1164.all;

entity Turnstile_Example is
  port (
    i_Reset  : in std_logic;
    i_Clk    : in std_logic;
    i_Coin   : in std_logic;
    i_Push   : in std_logic;
    o_Locked : out std_logic);
end entity Turnstile_Example;

architecture RTL of Turnstile_Example is

❶ type t_State is (LOCKED, UNLOCKED);
   signal r_Curr_State, r_Next_State : t_State;

begin

  -- Current state register
❷ process (i_Clk, i_Reset) is
  begin
    if i_Reset = '1' then
    ❸ r_Curr_State <= LOCKED;
    elsif rising_edge(i_Clk) then
    ❹ r_Curr_State <= r_Next_State;
    end if;
  end process;

  -- Next state determination
❺ process (r_Curr_State, i_Coin, i_Push)
  begin
    r_Next_State <= r_Curr_State;

  ❻ case r_Curr_State is

      when LOCKED =>
        if i_Coin = '1' then
        ❼ r_Next_State <= UNLOCKED;
        end if;

      when UNLOCKED =>
        if i_Push = '1' then
```

```
            r_Next_State <= LOCKED;
        end if;
    ❽
    end case;
end process;

❾ o_Locked <= '1' when r_Curr_State = LOCKED else '0';

end RTL;
```

We create the states using enumeration ❶, meaning each state name
has an assigned number. In Verilog, you need to create the list of states
manually. I like to use localparam to define each state, assigning them numbers in an incrementing order. In VHDL, you instead create a user-defined
type for your state machine (t_State). Then you list the states in order, and
VHDL automatically assigns them numbers. If you're familiar with C programming, the VHDL method is similar to how enumerations work in C.

NOTE *SystemVerilog supports automatic enumeration, but it doesn't exist in regular Verilog,
so we just number the states manually in the Verilog code.*

The first always or process block ❷, the current state register, is driven
by a clock. It keeps track of the current state by assigning r_Next_State to
r_Curr_State on every rising clock edge ❹. Notice that this block also has
the i_Reset signal in its sensitivity list, and that this signal is checked in
the block. It's important to include a way to get the state machine to its
initial condition, and we use i_Reset for that. The block's if...else statement (Verilog) or if...elsif statement (VHDL) checks whether i_Reset is
high *before* checking to see if we have a rising edge of the clock. This means
we're using an *asynchronous reset*; the reset can occur at any time, not necessarily on the rising edge of the clock. When i_Reset is high, we set the current state to LOCKED ❸. This is in keeping with the initial state indication in
Figure 8-1.

The second always or process block ❺ is the combinational one. It contains the logic for determining how to set r_Next_State. Notice that the sensitivity list ❺ and the block itself don't include a clock, so this block won't
generate any flip-flops, just LUTs. We set r_Next_State using a case statement
tied to the current state ❻, and by looking at our inputs. For example, if the
state is currently LOCKED and the i_Coin input is high, then the next state will
be UNLOCKED ❼. Compare the case statement and conditional logic with the
state machine diagram in Figure 8-1, and you'll see that we've addressed
all the transitions and events shown in the diagram that result in an actual
change of state. We don't need to write code for transitions that don't cause
the signal to change. For example, if the current state is LOCKED and i_Push
is high, we'll just stay in the LOCKED state. We could add a check for i_Push
in the LOCKED case and write r_Next_State <= LOCKED to make this explicit, but
that's unnecessary. Adding this line can make the designer's intent clearer,
but it also clutters up the code with additional assignments. It's up to you
which style you prefer.

We could also add a default case (in Verilog) or a when others case (in VHDL) above the endcase or end case statement ❽, to cover any conditions not explicitly called out in the state machine. Again, this is not required, but it can be a good idea; if you forget or omit a case, the default case will catch it. In this instance, I chose not to include a default. In fact, my code editor displays a suggestion for the VHDL when I try to include it:

Case statement contains all choices explicitly. You can safely remove the redundant 'others'(13).

The code ends by assigning the module's single output, o_Locked ❾. It will be high when we're in the LOCKED state, or low otherwise. If this code were really controlling a physical turnstile, we'd use changes in this output to trigger the actions that occur during state transitions, such as enabling or disabling the mechanism that locks the turnstile.

Using One always or process Block

The other approach to implementing a state machine combines all the logic into a single always or process block. As synthesis tools have improved over the years, they've gotten much better at understanding when you want to create a state machine, and where this one-block approach may once have been hard to synthesize, it's now perfectly viable (and arguably more straightforward to code). Here's the same turnstile state machine implemented with a single always or process block:

Verilog
```
module Turnstile_Example
 (input i_Reset,
  input i_Clk,
  input i_Coin,
  input i_Push,
  output o_Locked);

  localparam LOCKED   = 1'b0;
  localparam UNLOCKED = 1'b1;

  reg r_Curr_State;

  // Single always block approach
❶ always @(posedge i_Clk or posedge i_Reset)
  begin
    if (i_Reset)
      r_Curr_State <= LOCKED;
    else
    begin
❷ case (r_Curr_State)

      LOCKED:
        if (i_Coin)
          r_Curr_State <= UNLOCKED;
```

```
        UNLOCKED:
          if (i_Push)
            r_Curr_State <= LOCKED;

        endcase
      end
    end

    assign o_Locked = (r_Curr_State == LOCKED);

endmodule
```

```
library ieee;
use ieee.std_logic_1164.all;

entity Turnstile_Example is
  port (
    i_Reset  : in std_logic;
    i_Clk    : in std_logic;
    i_Coin   : in std_logic;
    i_Push   : in std_logic;
    o_Locked : out std_logic);
end entity Turnstile_Example;

architecture RTL of Turnstile_Example is

  type t_State is (LOCKED, UNLOCKED);
  signal r_Curr_State : t_State;

begin

  -- Single always block approach
❶ process (i_Clk, i_Reset) is
  begin
    if (i_Reset) then
      r_Curr_State <= LOCKED;
    elsif rising_edge(i_Clk) then

❷   case r_Curr_State is

      when LOCKED =>
        if i_Coin = '1' then
          r_Curr_State <= UNLOCKED;
        end if;

      when UNLOCKED =>
        if i_Push = '1' then
          r_Curr_State <= LOCKED;
        end if;

    end case;
  end if;
end process;
```

```
    o_Locked <= '1' when r_Curr_State = LOCKED else '0';

end RTL;
```

Everything works the same in the two approaches; the differences are purely stylistic, not functional. In this version of the state machine, we have a single always or process block ❶ that's sensitive to the clock and the reset signal. Rather than having both r_Curr_State and r_Next_State to worry about, we now only have r_Curr_State. None of the actual logic has changed, however. All we've done is move the work that was being done in the combinational always or process block into the sequential one, so the case statement will be evaluated at every rising clock edge ❷.

I'm not a big fan of the first approach we looked at, with the two always or process blocks. There are a few reasons for this. First, separating the LUT-based logic and the flip-flop-based logic into two separate blocks can be confusing, especially for beginners. Compared to the single-block solution, the design is more complicated and less intuitive, and it's easier to make mistakes. Second, as I said back in Chapter 4, I prefer not to use combinational-only always or process blocks if I can avoid them. They can generate latches if you're not careful, which can result in unwanted behavior. I recommend keeping your state machine logic within a single always or process block. The code is easier to read and understand, and the tools are good enough now to build the state machine correctly.

Testing the Design

Let's generate a testbench for this state machine to ensure that we're getting the output we desire:

Verilog
```
module Turnstile_Example_TB();

❶ reg r_Reset = 1'b1, r_Clk = 1'b0, r_Coin = 1'b0, r_Push = 1'b0;
  wire w_Locked;

  Turnstile_Example UUT
   (.i_Reset(r_Reset),
    .i_Clk(r_Clk),
    .i_Coin(r_Coin),
    .i_Push(r_Push),
    .o_Locked(w_Locked));

  always #1 r_Clk <= !r_Clk;

  initial begin
    $dumpfile("dump.vcd");
    $dumpvars;
    #10;
❷  r_Reset <= 1'b0;
    #10;
❸  assert (w_Locked == 1'b1);
```

```verilog
❹ r_Coin <= 1'b1;
    #10;
    assert (w_Locked == 1'b0);

    r_Push <= 1'b1;
    #10;
    assert (w_Locked == 1'b1);

    r_Coin <= 1'b0;
    #10;
    assert (w_Locked == 1'b1);

    r_Push <= 1'b0;
    #10;
    assert (w_Locked == 1'b1);
    $finish();
  end

endmodule
```

VHDL
```vhdl
library ieee;
use ieee.std_logic_1164.all;
use std.env.finish;

entity Turnstile_Example_TB is
end entity Turnstile_Example_TB;

architecture test of Turnstile_Example_TB is

❶ signal r_Reset : std_logic := '1';
  signal r_Clk, r_Coin, r_Push : std_logic := '0';
  signal w_Locked : std_logic;

begin

  UUT : entity work.Turnstile_Example
  port map (
    i_Reset  => r_Reset,
    i_Clk    => r_Clk,
    i_Coin   => r_Coin,
    i_Push   => r_Push,
    o_Locked => w_Locked);

  r_Clk <= not r_Clk after 1 ns;

  process is
  begin
    wait for 10 ns;
❷ r_Reset <= '0';
    wait for 10 ns;
❸ assert w_Locked = '1' severity failure;
```

```
❹ r_Coin <= '1';
  wait for 10 ns;
  assert w_Locked = '0' severity failure;

  r_Push <= '1';
  wait for 10 ns;
  assert w_Locked = '1' severity failure;

  r_Coin <= '0';
  wait for 10 ns;
  assert w_Locked = '1' severity failure;

  r_Push <= '0';
  wait for 10 ns;
  assert w_Locked = '1' severity failure;

  finish;  -- need VHDL-2008
end process;

end test;
```

This testbench drives the inputs in all possible combinations and monitors the single output (w_Locked) to see how it behaves. For example, r_Reset is initialized to high at the outset ❶, which should put us in the LOCKED state. Then, after 10 ns, we drive r_Reset low ❷. This should have no effect on the state, so we use the assert keyword in both VHDL and Verilog to verify that we're in the LOCKED state (indicated by a w_Locked value of 1) ❸. We then continue manipulating the other inputs and asserting the expected output (for instance, driving r_Coin high ❹ should put us in an UNLOCKED state). Our use of assert statements to automatically alert us to any failures makes this a self-checking testbench.

NOTE *For Verilog users, remember that* assert *only exists in SystemVerilog. Be sure to tell the simulator that your testbench is a SystemVerilog file, rather than regular Verilog.*

State Machine Best Practices

Before moving on, I want to share some recommendations for developing successful state machines. These are guidelines that I find helpful when I write my own FPGA state machines, and they're all modeled in the turnstile example we reviewed in the previous section:

Include one state machine per file.

It's certainly possible to write many state machines within a single file, but I strongly suggest that you limit the scope of any given Verilog or VHDL file to a single state machine. When you put two or more state machines in the same file, it can be hard to keep them from getting logically intertwined. It might require more typing to break things into multiple files, but it will save you time in the debugging stage.

Use the single-block approach.

As I've stated, I find that it's easier to write cleaner and less error-prone state machines if you only have one always or process block to worry about, rather than two. The one-block approach also avoids the need for combinational always or process blocks, which can generate latches if you're not careful.

Give your states meaningful names.

It's much easier to read a case statement that has actual words associated with each of the cases, provided you've named your states thoughtfully. For example, use descriptive names like IDLE, START_COUNT, LOCKED, UNLOCKED, and so on, rather than generic names like S0, S1, S2, and S3. Meaningful state names will help other people reading the code understand what's going on. Additionally, you'll thank yourself for descriptive state naming when you come back to your code after not looking at it for months. Enumeration allows you to do this. Enumeration is a common programming technique that allows you to use words in place of integers in your code. This is done either through localparam in Verilog or a user-defined type in VHDL.

Draw your state machine flow before coding.

Diving headfirst into coding a state machine is a recipe for disaster. Begin by drawing a diagram of the state machine that you want to implement, like the one you saw in Figure 8-1. This will help you ensure that you've thought through the entire flow, from the initial transition through all the possible permutations. If you realize that you've missed something once you start working on the code, that's fine; just be sure to go back and update your diagram to keep it in sync with the code you're writing. Your future self will thank you if you have this documentation.

It is by no means mandatory to follow these suggestions, but doing so will help you avoid some common pitfalls and create state machines within your FPGAs that are bug-free, easy to understand, and easy to maintain. The more complex your state machines become, the more helpful these best practices will be.

Project #6: Creating a Memory Game

We'll now put what you've learned about state machines into action by creating a memory game that runs on your development board. The player will have to remember and reproduce a pattern that grows longer as the game progresses, similar to a game like Simon. If the player can remember the entire pattern, they win.

The pattern is displayed using four LEDs. It starts simply, with just one LED lighting up. Then it's the player's turn to re-create the pattern by pressing the switch that matches the LED. If they push the wrong switch, the game is over. If they push the correct switch, the game continues, with the pattern expanding to a sequence of two LED blinks. The pattern keeps expanding until it's seven blinks long (although you'll be able to adjust the code to make it longer if you want). If the player re-creates the last pattern correctly, they may choose to play another round with a new pattern.

This project takes advantage of a peripheral that we haven't used before: a *seven-segment display*. This device uses an arrangement of seven LEDs to display the digits 0 through 9 (and a selection of letters), like something you'd see on a digital clock. It will serve as a scoreboard, keeping track of the player's progress through the pattern. We'll also use it to display an F (for *failure*) if the player makes a mistake, or an A when the game is won.

NOTE *If your development board doesn't have four LEDs and switches, you can adapt the project's code to work with the resources available. If it doesn't have a seven-segment display, try connecting one to your board, for example using a Pmod connector.*

Planning the State Machine

To create the game, we'll need to control the FPGA's flow through various states of operation, such as displaying the pattern and waiting for the player's response. Sounds like a perfect opportunity to use a state machine! In keeping with our best practices, we'll use a diagram to plan out the state machine before we write any code. Figure 8-3 shows a diagram for a state machine that satisfies the description of the game.

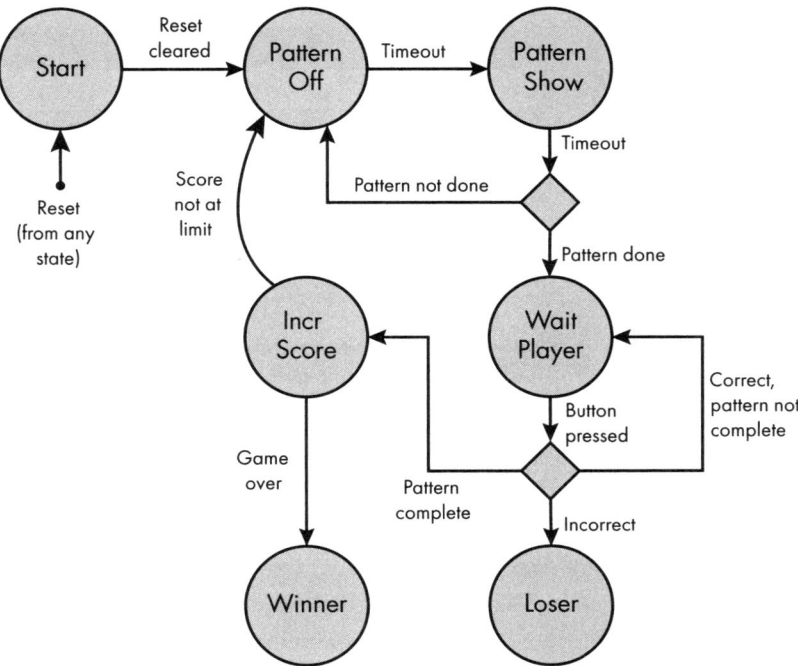

Figure 8-3: The memory game state machine diagram

Starting from the top-left corner, we have a reset/initial condition, which will jump into the Start state from any other state. I didn't draw arrows from every state back to Start, to avoid cluttering the diagram; just remember that you can always jump back to Start from any state when the reset condition occurs. We remain in the Start state until the reset is cleared, at which point we transition into the Pattern Off state. We wait here with all LEDs off for a set amount of time, and then we transition into the Pattern Show state, where we illuminate a single LED from the pattern, again for a set amount of time. If it's the last LED in the pattern (the pattern is done), we then transition to the Wait Player state to await the player's response. If the LED pattern is not done, we transition back to Pattern Off. We keep cycling between Pattern Show and Pattern Off, lighting up the LEDs in the pattern one at a time, until the pattern is done. The transitions back to Pattern Off add a pause between each blink, which avoids ambiguity in cases where the pattern includes the same LED twice in a row. This is the part of the game where the LED pattern is being shown to the player, for them to try to re-create later on.

NOTE *The diamond in the diagram between the Pattern Show and Wait Player states represents a* guard condition, *a Boolean expression that determines the state machine flow. In this case, the guard condition is checking whether the pattern is done.*

Once we're in the Wait Player state, the FPGA monitors the input from the buttons until one of two things happens. If the player pushes an incorrect button in the sequence, then we transition to the Loser state and show an F on the seven-segment display. If the player successfully re-creates the entire pattern, then we transition to the Incr Score (Increment Score) state. Here we check if the game is done, in which case the player has won and we transition to the Winner state, where we show an A on the seven-segment display. If the game isn't done, then we go back to Pattern Off to get ready to display the pattern again, this time with one additional LED blink added to the sequence.

There are nearly endless possibilities for designing state machines, so by no means is the arrangement shown in Figure 8-3 the only option. For example, we could have combined Pattern Off and Pattern Show into a single state that handles turning the LEDs both on and off. Our design, however, strikes a balance between the number of states and the complexity of each individual state. As a general rule, if one state is responsible for several actions, it might be an indication that the state should be broken into two or more states.

Organizing the Design

Next, we'll take a look at the overall organization of the project. Figure 8-4 shows a block diagram of the design.

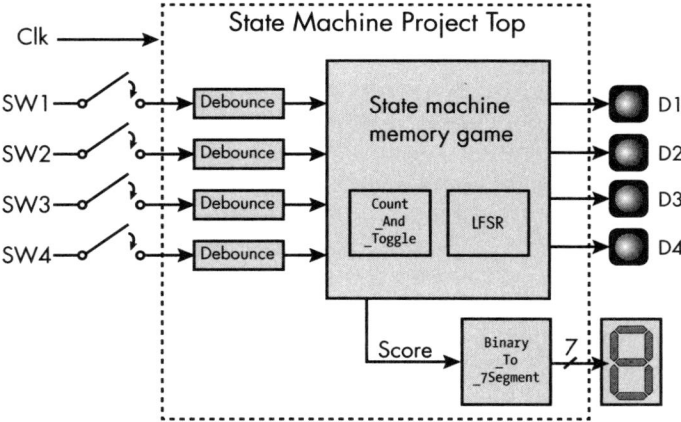

Figure 8-4: The Project #6 block diagram

Let's trace the flow of data through the block diagram. First, we have four switches (buttons) that are used to control the entire game. Remember that these are mechanical switches, so they're subject to bouncing. To get reliable button responses, these inputs must be debounced, which is the first thing that we do to each switch's signal as it enters the FPGA. We'll use the debounce filter module we implemented in Chapter 5 for this. The FPGA also has a clock input, which we'll use to drive all the flip-flops in this design.

Next, we have the memory game module itself, which is where the state machine lives. We'll explore this code in detail shortly. Notice that this module instantiates two submodules: `Count_And_Toggle` and the LFSR module, both of which you saw in Chapter 6. Remember that LFSRs are pseudorandom pattern generators, so we'll use one here to create a random pattern for the game. We'll use the `Count_And_Toggle` module to keep track of how long to display each LED in the pattern sequence; the toggling of this module will trigger transitions between states.

Finally, we have the `Binary_To_7Segment` module, which takes a binary input representing the player's score and drives the seven-segment display to light up that score. We'll look at how this works next.

Using the Seven-Segment Display

A seven-segment display consists of an arrangement of seven LEDs that can be lit in various combinations to produce different patterns. Figure 8-5 shows the seven segments of the display, labeled A through G. We'll use one of these displays to keep track of the score in this project, incrementing it each time the player successfully repeats the pattern.

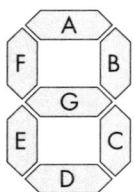

Figure 8-5:
A seven-segment
display

Conventionally, seven-segment displays are used to show the decimal numbers 0 through 9, but we can extend our display's range further by showing the hexadecimal numbers A through F (for 10 through 15) as well. We can't simply tell the display to light up a particular number, however, since each segment in the display is controlled separately. Instead, our Binary _To_7Segment module takes in the number to be shown and translates it into the appropriate signals for driving the display. Let's take a look at the code:

Verilog
```
module Binary_To_7Segment
 (input       i_Clk,
❶ input [3:0] i_Binary_Num,
  output      o_Segment_A,
  output      o_Segment_B,
  output      o_Segment_C,
  output      o_Segment_D,
  output      o_Segment_E,
  output      o_Segment_F,
  output      o_Segment_G);

  reg [6:0]    r_Hex_Encoding;

  always @(posedge i_Clk)
    begin
❷ case (i_Binary_Num)
       4'b0000 : r_Hex_Encoding <= 7'b1111110; // 0x7E
       4'b0001 : r_Hex_Encoding <= 7'b0110000; // 0x30
       4'b0010 : r_Hex_Encoding <= 7'b1101101; // 0x6D
       4'b0011 : r_Hex_Encoding <= 7'b1111001; // 0x79
       4'b0100 : r_Hex_Encoding <= 7'b0110011; // 0x33
       4'b0101 : r_Hex_Encoding <= 7'b1011011; // 0x5B
       4'b0110 : r_Hex_Encoding <= 7'b1011111; // 0x5F
❸ 4'b0111 : r_Hex_Encoding <= 7'b1110000; // 0x70
       4'b1000 : r_Hex_Encoding <= 7'b1111111; // 0x7F
       4'b1001 : r_Hex_Encoding <= 7'b1111011; // 0x7B
       4'b1010 : r_Hex_Encoding <= 7'b1110111; // 0x77
       4'b1011 : r_Hex_Encoding <= 7'b0011111; // 0x1F
       4'b1100 : r_Hex_Encoding <= 7'b1001110; // 0x4E
       4'b1101 : r_Hex_Encoding <= 7'b0111101; // 0x3D
       4'b1110 : r_Hex_Encoding <= 7'b1001111; // 0x4F
       4'b1111 : r_Hex_Encoding <= 7'b1000111; // 0x47
       default : r_Hex_Encoding <= 7'b0000000; // 0x00
```

```
          endcase
        end

❹ assign o_Segment_A = r_Hex_Encoding[6];
  assign o_Segment_B = r_Hex_Encoding[5];
  assign o_Segment_C = r_Hex_Encoding[4];
  assign o_Segment_D = r_Hex_Encoding[3];
  assign o_Segment_E = r_Hex_Encoding[2];
  assign o_Segment_F = r_Hex_Encoding[1];
  assign o_Segment_G = r_Hex_Encoding[0];


endmodule
```

VHDL
```
library ieee;
use ieee.std_logic_1164.all;

entity Binary_To_7Segment is
  port (
    i_Clk        : in std_logic;
❶ i_Binary_Num : in std_logic_vector(3 downto 0);
    o_Segment_A  : out std_logic;
    o_Segment_B  : out std_logic;
    o_Segment_C  : out std_logic;
    o_Segment_D  : out std_logic;
    o_Segment_E  : out std_logic;
    o_Segment_F  : out std_logic;
    o_Segment_G  : out std_logic
    );
end entity Binary_To_7Segment;

architecture RTL of Binary_To_7Segment is
  signal r_Hex_Encoding : std_logic_vector(6 downto 0);
begin

  process (i_Clk) is
  begin
    if rising_edge(i_Clk) then
    ❷ case i_Binary_Num is
        when "0000" =>
          r_Hex_Encoding <= "1111110"; -- 0x7E
        when "0001" =>
          r_Hex_Encoding <= "0110000"; -- 0x30
        when "0010" =>
          r_Hex_Encoding <= "1101101"; -- 0x6D
        when "0011" =>
          r_Hex_Encoding <= "1111001"; -- 0x79
        when "0100" =>
          r_Hex_Encoding <= "0110011"; -- 0x33
        when "0101" =>
          r_Hex_Encoding <= "1011011"; -- 0x5B
        when "0110" =>
          r_Hex_Encoding <= "1011111"; -- 0x5F
      ❸ when "0111" =>
          r_Hex_Encoding <= "1110000"; -- 0x70
```

```
          when "1000" =>
              r_Hex_Encoding <= "1111111"; -- 0x7F
          when "1001" =>
              r_Hex_Encoding <= "1111011"; -- 0x7B
          when "1010" =>
              r_Hex_Encoding <= "1110111"; -- 0x77
          when "1011" =>
              r_Hex_Encoding <= "0011111"; -- 0x1F
          when "1100" =>
              r_Hex_Encoding <= "1001110"; -- 0x4E
          when "1101" =>
              r_Hex_Encoding <= "0111101"; -- 0x3D
          when "1110" =>
              r_Hex_Encoding <= "1001111"; -- 0x4F
          when "1111" =>
              r_Hex_Encoding <= "1000111"; -- 0x47
          when others =>
              r_Hex_Encoding <= "0000000"; -- 0x00
        end case;
      end if;
    end process;

❹ o_Segment_A <= r_Hex_Encoding(6);
  o_Segment_B <= r_Hex_Encoding(5);
  o_Segment_C <= r_Hex_Encoding(4);
  o_Segment_D <= r_Hex_Encoding(3);
  o_Segment_E <= r_Hex_Encoding(2);
  o_Segment_F <= r_Hex_Encoding(1);
  o_Segment_G <= r_Hex_Encoding(0);

end architecture RTL;
```

This module takes a 4-bit binary input ❶ and uses seven outputs to light up the appropriate segments in the display, given the input. The case statement ❷ captures all possible inputs, from 0000 to 1111 (0 through 15), and translates each number into the correct output pattern using the 7-bit r_Hex_Encoding register. Each bit in the register maps to one of the segments in the display: bit 6 maps to segment A, bit 5 maps to segment B, and so on. To see how this works, let's consider a specific input—say, 0111, which is the digit 7—as an example. Figure 8-6 illustrates how to illuminate a seven-segment display to show this digit.

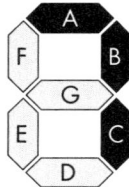

Figure 8-6:
Illuminating a
7 on a seven-
segment display

As you can see in the figure, we need to illuminate segments A, B, and C, while keeping the other segments turned off, to show the digit 7. In the code at ❸, we therefore set r_Hex_Encoding to 0x70, or 1110000 in binary, putting 1s on the three bits corresponding to segments A, B, and C. Then, outside the case statement, we extract each individual bit from the register and pass it to the appropriate output using a continuous assignment ❹. This approach of encoding the pattern into the r_Hex_Encoding register saves a lot of typing; we don't have to assign all seven outputs in every single branch of the case statement.

Coding the Top-Level Module

Next, let's jump into the top module of the project to see how everything is wired up at the highest level. If you refer back to the block diagram in Figure 8-4, you'll see this module represented by the square with the dotted line:

Verilog

```verilog
module State_Machine_Project_Top
 (input i_Clk,
  // Input switches for entering pattern
  input i_Switch_1,
  input i_Switch_2,
  input i_Switch_3,
  input i_Switch_4,
  // Output LEDs for displaying pattern
  output o_LED_1,
  output o_LED_2,
  output o_LED_3,
  output o_LED_4,
  // Scoreboard, 7-segment display
  output o_Segment2_A,
  output o_Segment2_B,
  output o_Segment2_C,
  output o_Segment2_D,
  output o_Segment2_E,
  output o_Segment2_F,
  output o_Segment2_G);

❶ localparam GAME_LIMIT    = 7;        // Increase to make game harder
  localparam CLKS_PER_SEC  = 25000000; // 25 MHz clock
  localparam DEBOUNCE_LIMIT = 250000;  // 10 ms debounce filter

  wire w_Switch_1, w_Switch_2, w_Switch_3, w_Switch_4;
  wire w_Segment2_A, w_Segment2_B, w_Segment2_C, w_Segment2_D;
  wire w_Segment2_E, w_Segment2_F, w_Segment2_G;
  wire [3:0] w_Score;

  // Debounce all switch inputs to remove mechanical glitches
❷ Debounce_Filter #(.DEBOUNCE_LIMIT(DEBOUNCE_LIMIT)) Debounce_SW1
   (.i_Clk(i_Clk),
    .i_Bouncy(i_Switch_1),
    .o_Debounced(w_Switch_1));
```

```verilog
    Debounce_Filter #(.DEBOUNCE_LIMIT(DEBOUNCE_LIMIT)) Debounce_SW2
     (.i_Clk(i_Clk),
      .i_Bouncy(i_Switch_2),
      .o_Debounced(w_Switch_2));

    Debounce_Filter #(.DEBOUNCE_LIMIT(DEBOUNCE_LIMIT)) Debounce_SW3
     (.i_Clk(i_Clk),
      .i_Bouncy(i_Switch_3),
      .o_Debounced(w_Switch_3));

    Debounce_Filter #(.DEBOUNCE_LIMIT(DEBOUNCE_LIMIT)) Debounce_SW4
     (.i_Clk(i_Clk),
      .i_Bouncy(i_Switch_4),
      .o_Debounced(w_Switch_4));

❸ State_Machine_Game #(.CLKS_PER_SEC(CLKS_PER_SEC),
                       .GAME_LIMIT(GAME_LIMIT)) Game_Inst
     (.i_Clk(i_Clk),
      .i_Switch_1(w_Switch_1),
      .i_Switch_2(w_Switch_2),
      .i_Switch_3(w_Switch_3),
      .i_Switch_4(w_Switch_4),
      .o_Score(w_Score),
      .o_LED_1(o_LED_1),
      .o_LED_2(o_LED_2),
      .o_LED_3(o_LED_3),
      .o_LED_4(o_LED_4));

❹ Binary_To_7Segment Scoreboard
     (.i_Clk(i_Clk),
      .i_Binary_Num(w_Score),
      .o_Segment_A(w_Segment2_A),
      .o_Segment_B(w_Segment2_B),
      .o_Segment_C(w_Segment2_C),
      .o_Segment_D(w_Segment2_D),
      .o_Segment_E(w_Segment2_E),
      .o_Segment_F(w_Segment2_F),
      .o_Segment_G(w_Segment2_G));

❺ assign o_Segment2_A = !w_Segment2_A;
  assign o_Segment2_B = !w_Segment2_B;
  assign o_Segment2_C = !w_Segment2_C;
  assign o_Segment2_D = !w_Segment2_D;
  assign o_Segment2_E = !w_Segment2_E;
  assign o_Segment2_F = !w_Segment2_F;
  assign o_Segment2_G = !w_Segment2_G;

endmodule
```

VHDL
```vhdl
library IEEE;
use IEEE.std_logic_1164.all;

entity State_Machine_Project_Top is
```

```
  port (
    i_Clk : in std_logic;
    -- Input switches for entering pattern
    i_Switch_1 : in std_logic;
    i_Switch_2 : in std_logic;
    i_Switch_3 : in std_logic;
    i_Switch_4 : in std_logic;
    -- Output LEDs for displaying pattern
    o_LED_1 : out std_logic;
    o_LED_2 : out std_logic;
    o_LED_3 : out std_logic;
    o_LED_4 : out std_logic;
    -- Scoreboard, 7-segment display
    o_Segment2_A : out std_logic;
    o_Segment2_B : out std_logic;
    o_Segment2_C : out std_logic;
    o_Segment2_D : out std_logic;
    o_Segment2_E : out std_logic;
    o_Segment2_F : out std_logic;
    o_Segment2_G : out std_logic);

end entity State_Machine_Project_Top;

architecture RTL of State_Machine_Project_Top is

❶ constant GAME_LIMIT     : integer := 7;        -- Increase to make game harder
  constant CLKS_PER_SEC   : integer := 25000000; -- 25 MHz clock
  constant DEBOUNCE_LIMIT : integer := 250000;   -- 10 ms debounce filter

  signal w_Switch_1, w_Switch_2, w_Switch_3, w_Switch_4 : std_logic;
  signal w_Score : std_logic_vector(3 downto 0);
  signal w_Segment2_A, w_Segment2_B, w_Segment2_C, w_Segment2_D : std_logic;
  signal w_Segment2_E, w_Segment2_F, w_Segment2_G : std_logic;

begin

❷ Debounce_SW1 : entity work.Debounce_Filter
    generic map (
      DEBOUNCE_LIMIT => DEBOUNCE_LIMIT)
    port map (
      i_Clk       => i_Clk,
      i_Bouncy    => i_Switch_1,
      o_Debounced => w_Switch_1);

  Debounce_SW2 : entity work.Debounce_Filter
    generic map (
      DEBOUNCE_LIMIT => DEBOUNCE_LIMIT)
    port map (
      i_Clk       => i_Clk,
      i_Bouncy    => i_Switch_2,
      o_Debounced => w_Switch_2);

  Debounce_SW3 : entity work.Debounce_Filter
    generic map (
      DEBOUNCE_LIMIT => DEBOUNCE_LIMIT)
```

```
      port map (
        i_Clk        => i_Clk,
        i_Bouncy     => i_Switch_3,
        o_Debounced => w_Switch_3);

    Debounce_SW4 : entity work.Debounce_Filter
      generic map (
        DEBOUNCE_LIMIT => DEBOUNCE_LIMIT)
      port map (
        i_Clk => i_Clk,
        i_Bouncy => i_Switch_4,
        o_Debounced => w_Switch_4);

❸ Game_Inst : entity work.State_Machine_Game
    generic map (
      CLKS_PER_SEC => CLKS_PER_SEC,
      GAME_LIMIT   => GAME_LIMIT)
    port map (
      i_Clk        => i_Clk,
      i_Switch_1 => w_Switch_1,
      i_Switch_2 => w_Switch_2,
      i_Switch_3 => w_Switch_3,
      i_Switch_4 => w_Switch_4,
      o_Score     => w_Score,
      o_LED_1     => o_LED_1,
      o_LED_2     => o_LED_2,
      o_LED_3     => o_LED_3,
      o_LED_4     => o_LED_4);

❹ Scoreboard : entity work.Binary_To_7Segment
    port map (
      i_Clk        => i_Clk,
      i_Binary_Num => w_Score,
      o_Segment_A  => w_Segment2_A,
      o_Segment_B  => w_Segment2_B,
      o_Segment_C  => w_Segment2_C,
      o_Segment_D  => w_Segment2_D,
      o_Segment_E  => w_Segment2_E,
      o_Segment_F  => w_Segment2_F,
      o_Segment_G  => w_Segment2_G);

❺ o_Segment2_A <= not w_Segment2_A;
  o_Segment2_B <= not w_Segment2_B;
  o_Segment2_C <= not w_Segment2_C;
  o_Segment2_D <= not w_Segment2_D;
  o_Segment2_E <= not w_Segment2_E;
  o_Segment2_F <= not w_Segment2_F;
  o_Segment2_G <= not w_Segment2_G;

end RTL;
```

My goal for writing the top module of a design, especially as the design becomes more complicated, is to minimize the amount of functional code within it. Ideally, code that performs functionality should be pushed into lower levels, so the highest level is just wires and module instantiation. This helps keep the code clean and ensures that each module is focused on performing what's needed of it, without spreading the functionality across multiple layers.

For this project, we first instantiate four debounce filter modules, one for each push button ❷. Then we instantiate the State_Machine_Game module, which contains the logic for the state machine and the game itself ❸. The inputs to this module, w_Switch_1 through w_Switch_4, are the outputs of the debounce filters, so this module can trust that the input signals are stable. Notice that the module has two parameters (Verilog) or generics (VHDL), CLKS_PER_SEC and GAME_LIMIT, both of which were set earlier ❶. The former specifies the number of clock cycles per second (needed for keeping track of time), and is there in case the design is run at a different clock frequency. The latter controls the maximum length of the pattern.

Next we instantiate the Binary_To_7Segment module ❹, which takes the w_Score output from the game as an input so the score will be displayed to the player. Notice, however, that we invert all the outputs from the display module before outputting them at the top level ❺. A low on the output may be needed, rather than a high, to light up each segment, depending on the way the seven-segment display is connected on your development board's PCB. If your display isn't behaving as expected, try removing the !s from the Verilog or nots from the VHDL to avoid inverting the outputs.

Our top-level module doesn't instantiate the LFSR or Count_And_Toggle modules directly: those are instantiated within the State_Machine_Game module. You're starting to see here how a hierarchy can be established inside an FPGA, and how a complex design can be built up from relatively simple modules.

Coding the State Machine

Now let's get to the meat of the project: the state machine itself. We'll examine the State_Machine_Game module in sections, but remember that you can view the complete code listing in the book's GitHub repository (*https://github.com/nandland/getting-started-with-fpgas*). The module starts, as usual, by declaring the inputs, outputs, and internal signals:

Verilog

```
module State_Machine_Game # (parameter CLKS_PER_SEC = 25000000,
                             parameter GAME_LIMIT = 6)
 (input i_Clk,
  input i_Switch_1,
  input i_Switch_2,
  input i_Switch_3,
  input i_Switch_4,
  output reg [3:0] o_Score,
  output o_LED_1,
  output o_LED_2,
  output o_LED_3,
  output o_LED_4
  );
```

```
❶ localparam START       = 3'd0;
  localparam PATTERN_OFF  = 3'd1;
  localparam PATTERN_SHOW = 3'd2;
  localparam WAIT_PLAYER  = 3'd3;
  localparam INCR_SCORE   = 3'd4;
  localparam LOSER        = 3'd5;
  localparam WINNER       = 3'd6;

❷ reg [2:0] r_SM_Main;
  reg r_Toggle, r_Switch_1, r_Switch_2, r_Switch_3;
  reg r_Switch_4, r_Button_DV;
❸ reg [1:0] r_Pattern[0:10]; // 2D array: 2 bits wide x 11 deep
  wire [21:0] w_LFSR_Data;
  reg [$clog2(GAME_LIMIT)-1:0] r_Index; // Display index
  reg [1:0] r_Button_ID;
  wire w_Count_En, w_Toggle;
--snip--
```

```
VHDL  library IEEE;
      use IEEE.std_logic_1164.all;
      use IEEE.numeric_std.all;

      entity State_Machine_Game is
        generic (
          CLKS_PER_SEC : integer := 25000000;
          GAME_LIMIT   : integer := 6);
        port(
          i_Clk      : in std_logic;
          i_Switch_1 : in std_logic;
          i_Switch_2 : in std_logic;
          i_Switch_3 : in std_logic;
          i_Switch_4 : in std_logic;
          o_Score    : out std_logic_vector(3 downto 0);
          o_LED_1    : out std_logic;
          o_LED_2    : out std_logic;
          o_LED_3    : out std_logic;
          o_LED_4    : out std_logic);

      end entity State_Machine_Game;

      architecture RTL of State_Machine_Game is

    ❶ type t_SM_Main is (START, PATTERN_OFF, PATTERN_SHOW,
                         WAIT_PLAYER, INCR_SCORE, LOSER, WINNER);

    ❷ signal r_SM_Main : t_SM_Main;
      signal w_Count_En, w_Toggle, r_Toggle, r_Switch_1 : std_logic;
      signal r_Switch_2, r_Switch_3, r_Switch_4, r_Button_DV : std_logic;

      type t_Pattern is array (0 to 10) of std_logic_vector(1 downto C);
    ❸ signal r_Pattern : t_Pattern; -- 2D Array: 2-bit wide x 11 deep
```

```
signal w_LFSR_Data : std_logic_vector(21 downto 0);
signal r_Index : integer range 0 to GAME_LIMIT;
signal w_Index_SLV : std_logic_vector(7 downto 0);
signal r_Button_ID : std_logic_vector(1 downto 0);
signal r_Score : unsigned(3 downto 0);
--snip--
```

We use the enumeration approach described earlier in the chapter to name each state ❶. The r_SM_Main signal ❷ will keep track of the current state. It needs to have enough bits to convey all the possible states. In this case, we have seven total states, which can fit inside a 3-bit-wide register. In Verilog, we explicitly declare the signal as having 3 bits. In VHDL, however, we just create the state machine signal to be of the custom t_SM_Main data type (the enumeration that we created ❶), and it will be sized automatically.

Another important signal we're creating is r_Pattern, which stores the pattern for the game ❸. This is the second time that we've created a two-dimensional signal within our FPGA (the first time was back in Chapter 6, when we were creating RAM). Specifically, r_Pattern is 2 bits wide by 11 items deep, for a total storage of 22 bits. Each pair of bits in this signal corresponds to one of the four LEDs (00 indicates LED1, 01 indicates LED2, and so on), giving us a sequence of LEDs to light up (and a sequence of switches to press). Table 8-1 shows an example of what the data in this 2D register might look like.

Table 8-1: Pattern Storage Example

Index	Binary	LED/switch
0	01	2
1	11	4
2	11	4
3	00	1
4	10	3
5	00	1
6	01	2
7	01	2
8	11	4
9	00	1
10	10	3

In this example, the value at index 0 is 01, which correlates to the second LED/switch, the value at index 1 is 11, which correlates to the fourth LED/switch, and so on. We'll be able to use the index to increment through the register, getting 2 bits for each index. The binary pattern itself comes from the LFSR, and will be random each time. The LFSR is 22 bits wide, so each bit of the LFSR output is mapped to a bit in this 2D register. This means that the maximum length of the memory pattern that we can create is 11 LED blinks long. After playing this game several times, however, I can tell you that it gets quite challenging to remember patterns that go that high. As I noted earlier, the actual limit for the game is set by the parameter/generic GAME_LIMIT, which can be overridden from the top module. If you want to set the game to the maximum difficulty, try changing GAME_LIMIT to 11.

The module continues by handling reset conditions:

Verilog

```
--snip--
  always @(posedge i_Clk)
  begin

    // Reset game from any state
❶ if (i_Switch_1 & i_Switch_2)
      r_SM_Main <= START;
    else
    begin

      // Main state machine switch statement
❷    case (r_SM_Main)
--snip--
```

VHDL

```
--snip--
begin

  process (i_Clk) is
  begin

    if rising_edge(i_Clk) then
      -- Reset game from any state
❶   if i_Switch_1 = '1' and i_Switch_2 = '1' then
        r_SM_Main <= START;
      else

        -- Main state machine switch statement
❷     case r_SM_Main is
--snip--
```

The player must push switch 1 and switch 2 at the same time to trigger the START state. We check for this with an if statement ❶. Notice that this check occurs outside the state machine's main case statement, which we initiate in the else branch ❷. This means that on every clock cycle we'll check if both switches are pressed, and enter the START state if they are or run the state machine for the game if they aren't. It would have been easier if we had a fifth button that was dedicated to resetting the state machine, but alas, I did not, so I had to be a bit creative here.

Now let's look at the first few states in the case statement:

Verilog

```
--snip--
    // Main state machine switch statement
    case (r_SM_Main)
```

```
     // Stay in START state until user releases buttons
❶ START:
   begin
     // Wait for reset condition to go away
  ❷ if (!i_Switch_1 & !i_Switch_2 & r_Button_DV)
     begin
       o_Score   <= 0;
       r_Index   <= 0;
       r_SM_Main <= PATTERN_OFF;
     end
   end

❸ PATTERN_OFF:
   begin
     if (!w_Toggle & r_Toggle) // Falling edge found
       r_SM_Main <= PATTERN_SHOW;
   end

   // Show the next LED in the pattern
❹ PATTERN_SHOW:
   begin
     if (!w_Toggle & r_Toggle) // Falling edge found
    ❺ if (o_Score == r_Index)
       begin
      ❻ r_Index   <= 0;
         r_SM_Main <= WAIT_PLAYER;
       end
       else
       begin
      ❼ r_Index   <= r_Index + 1;
         r_SM_Main <= PATTERN_OFF;
       end

   end
--snip--
```

VHDL --snip--
```
     -- Main state machine switch statement
     case r_SM_Main is

       -- Stay in START state until user releases buttons
❶ when START =>
     -- Wait for reset condition to go away
  ❷ if (i_Switch_1 = '0' and i_Switch_2 = '0' and
         r_Button_DV = '1') then
       r_Score   <= to_unsigned(0, r_Score'length);
       r_Index   <= 0;
       r_SM_Main <= PATTERN_OFF;
     end if;

❸ when PATTERN_OFF =>
     if w_Toggle = '0' and r_Toggle = '1' then -- Falling edge found
```

```
                    r_SM_Main <= PATTERN_SHOW;
                end if;

            -- Show the next LED in the pattern
        ❹ when PATTERN_SHOW =>
            if w_Toggle = '0' and r_Toggle = '1' then -- Falling edge found
            ❺ if r_Score = r_Index then
                ❻ r_Index    <= 0;
                  r_SM_Main <= WAIT_PLAYER;
                else
                ❼ r_Index    <= r_Index + 1;
                  r_SM_Main <= PATTERN_OFF;
                end if;
            end if;
--snip--
```

First we handle the START state ❶, where we wait for the reset condition to be removed. This happens when switches 1 and 2 are released ❷. Notice that we're looking not only for a low on the two switches, but also for a high on r_Button_DV. We use this signal throughout the module to detect falling edges—that is, releases—on the four switches. You'll see how this works later in the code. When the reset is cleared, we set the score and the pattern index to 0, then go into the PATTERN_OFF state.

The PATTERN_OFF state ❸ simply waits for a timer, driven by the Count_And _Toggle module, to expire. When this happens we transition to the PATTERN _SHOW state ❹, during which we'll be illuminating one of the LEDs (you'll see the code for illuminating the LEDs later). The transition out of PATTERN_SHOW is also triggered by the timer in the Count_And_Toggle module. When the timer expires we need to decide if we're done displaying the pattern, which we do by checking whether the player's score (o_Score) equals the current index into the pattern (r_Index) ❺. If it doesn't, we aren't done, so we increment r_Index to get ready to light up the next LED in the pattern ❼ and we go back to the PATTERN_OFF state. If we are done, we reset r_Index to 0 ❻ and transition to the WAIT_PLAYER state. Let's look at that now:

Verilog
```
--snip--
        WAIT_PLAYER:
        begin
        ❶ if (r_Button_DV)
            ❷ if (r_Pattern[r_Index] == r_Button_ID && r_Index == o_Score)
                begin
                  r_Index    <= 0;
                  r_SM_Main <= INCR_SCORE;
                end
            ❹ else if (r_Pattern[r_Index] != r_Button_ID)
                  r_SM_Main <= LOSER;
            ❺ else
                  r_Index <= r_Index + 1;
        end
--snip--
```

VHDL `--snip--`

```
          when WAIT_PLAYER =>
        ❶ if r_Button_DV = '1' then
          ❷ if (r_Pattern(r_Index) = r_Button_ID and
              ❸ unsigned(w_Index_SLV) = r_Score) then
              r_Index   <= 0;
              r_SM_Main <= INCR_SCORE;
          ❹ elsif r_Pattern(r_Index) /= r_Button_ID then
              r_SM_Main <= LOSER;
          ❺ else
              r_Index <= r_Index + 1;
            end if;
          end if;
```
`--snip--`

In the WAIT_PLAYER state, we wait for r_Button_DV to go high, indicating the player has pressed and released a switch ❶. Then we check if the player has correctly pressed the next switch in the pattern. As you'll see later, each time any switch is released, r_Button_ID is set to indicate which switch it was (00 for switch 1, 01 for switch 2, and so on), so we compare r_Button_ID with a value in r_Pattern, using r_Index as an index into the 2D array. There are three possibilities. If the switch is correct and we're at the end of the pattern ❷, we reset r_Index and transition to the INCR_SCORE state. If the switch is incorrect, we transition to the LOSER state ❹. Otherwise, the switch is correct but we aren't done with the pattern, so we increment r_Index and wait for the next press ❺. Notice in this last case that we don't explicitly assign the state, so r_SM_Main will just retain its previous assignment (WAIT_PLAYER). We could add a line at the end of the else statement that says r_SM_Main <= WAIT_PLAYER; but it's not necessary. If r_SM_Main isn't assigned, then we know that path doesn't cause a state change.

One difference between the Verilog and the VHDL is that in the latter we need to be very explicit about the types we're comparing. In the VHDL only, we need to cast w_Index_SLV to unsigned type ❸ so that we can compare it to r_Score, which is also of type unsigned. Verilog is much more forgiving, so we don't need this extra conversion. We'll discuss numerical data types in detail in Chapter 10.

Now let's look at the remaining states in the case statement:

Verilog `--snip--`

```
          // Used to increment score counter
        ❶ INCR_SCORE:
          begin
            o_Score <= o_Score + 1;
            if (o_Score == GAME_LIMIT-1)
              r_SM_Main <= WINNER;
            else
              r_SM_Main <= PATTERN_OFF;
          end
```

```verilog
      // Display 0xA on 7-segment display, wait for new game
    ❷ WINNER:
      begin
        o_Score <= 4'hA; // Winner!
      end

      // Display 0xF on 7-segment display, wait for new game
    ❸ LOSER:
      begin
        o_Score <= 4'hF; // Loser!
      end

    ❹ default:
        r_SM_Main <= START;
      endcase
    end

  end
--snip--
```

VHDL
```vhdl
--snip--
          -- Used to increment score counter
      ❶ when INCR_SCORE =>
          r_Score <= r_Score + 1;
          if r_Score = GAME_LIMIT then
            r_SM_Main <= WINNER;
          else
            r_SM_Main <= PATTERN_OFF;
          end if;

        -- Display 0xA on 7-segment display, wait for new game
      ❷ when WINNER =>
          r_Score <= X"A"; -- Winner!

        -- Display 0xF on 7-segment display, wait for new game
      ❸ when LOSER =>
          r_Score <= X"F"; -- Loser!

      ❹ when others =>
          r_SM_Main <= START;
        end case;
      end if;
    end if;
  end process;
--snip--
```

In INCR_SCORE ❶, we increment the score variable and compare it with GAME_LIMIT to check if the game is over. If so, we go to the WINNER state, and if not, we go back to PATTERN_OFF to continue the memory sequence. Notice that we'll only be in this state for a single clock cycle. You could perhaps make the argument that INCR_SCORE isn't a necessary state, and that this logic should happen in WAIT_PLAYER instead. I chose to treat INCR_SCORE as a separate state to avoid making WAIT_PLAYER too complicated.

For the WINNER ❷ and LOSER ❸ states, we simply set the score value to show an A or F on the seven-segment display and remain in the current state. The state machine can only leave these states in the event of a reset condition, in which switches 1 and 2 are both pressed at the same time.

We also include a default clause at the end of the case statement ❹, which specifies what behavior to take in the event that r_SM_Main isn't one of the previously defined states. This shouldn't ever happen, but it's good practice to create a default case where we go back to START. The end statements at the end of this listing close out the case statement, if...else statement, and always or process block that the state machine was wrapped in.

We've now finished coding the state machine itself. The rest of the code in the module handles logic for helping with the tasks that occur during the various states. First we have the code for randomly generating the pattern:

Verilog
```
--snip--
    // Register in the LFSR to r_Pattern when game starts
    // Each 2 bits of LFSR is one value for r_Pattern 2D array
    always @(posedge i_Clk)
    begin
❶   if (r_SM_Main == START)
      begin
        r_Pattern[0]  <= w_LFSR_Data[1:0];
        r_Pattern[1]  <= w_LFSR_Data[3:2];
        r_Pattern[2]  <= w_LFSR_Data[5:4];
        r_Pattern[3]  <= w_LFSR_Data[7:6];
        r_Pattern[4]  <= w_LFSR_Data[9:8];
        r_Pattern[5]  <= w_LFSR_Data[11:10];
        r_Pattern[6]  <= w_LFSR_Data[13:12];
        r_Pattern[7]  <= w_LFSR_Data[15:14];
        r_Pattern[8]  <= w_LFSR_Data[17:16];
        r_Pattern[9]  <= w_LFSR_Data[19:18];
        r_Pattern[10] <= w_LFSR_Data[21:20];
      end
    end
--snip--
```

VHDL
```
--snip--
    -- Register in the LFSR to r_Pattern when game starts
    -- Each 2 bits of LFSR is one value for r_Pattern 2D array
    process (i_Clk) is
    begin
      if rising_edge(i_Clk) then
❶     if r_SM_Main = START then
          r_Pattern(0)  <= w_LFSR_Data(1 downto 0);
          r_Pattern(1)  <= w_LFSR_Data(3 downto 2);
          r_Pattern(2)  <= w_LFSR_Data(5 downto 4);
          r_Pattern(3)  <= w_LFSR_Data(7 downto 6);
          r_Pattern(4)  <= w_LFSR_Data(9 downto 8);
          r_Pattern(5)  <= w_LFSR_Data(11 downto 10);
          r_Pattern(6)  <= w_LFSR_Data(13 downto 12);
          r_Pattern(7)  <= w_LFSR_Data(15 downto 14);
          r_Pattern(8)  <= w_LFSR_Data(17 downto 16);
```

```
            r_Pattern(9)  <= w_LFSR_Data(19 downto 18);
            r_Pattern(10) <= w_LFSR_Data(21 downto 20);
          end if;
        end if;
      end process;

    ❷ w_Index_SLV <= std_logic_vector(to_unsigned(r_Index, w_Index_SLV'length));
    --snip--
```

We need to generate a different pattern each time the game is played, while also making sure the pattern gets "locked in" once the game is underway. To do this, we first check if we're in the START state ❶. If so, the game isn't currently in progress, so we use the LFSR to create a new pattern. Recall that the output of our LFSR is a pseudorandom string of bits, which changes on every clock cycle. We take 2-bit sections from the LFSR output and place them into the 11 slots in r_Pattern. This will keep happening every clock cycle until the player releases switches 1 and 2, triggering the transition out of START. At that point, the current values of r_Pattern will be locked in for the duration of the game.

In VHDL, we also need to create an intermediary signal, w_Index_SLV ❷, which is just the std_logic_vector representation of r_Index. Again, since VHDL is strongly typed, you'll often see intermediary signals used for generating the "correct" signal types. I could have put this line anywhere since it's a combinational assignment; as long as it's outside of a process block, its precise location in the file makes no functional difference.

Next comes the code for illuminating the four LEDs:

Verilog
```
--snip--
  assign o_LED_1 = (r_SM_Main == PATTERN_SHOW &&
                    r_Pattern[r_Index] == 2'b00) ? 1'b1 : i_Switch_1;
  assign o_LED_2 = (r_SM_Main == PATTERN_SHOW &&
                    r_Pattern[r_Index] == 2'b01) ? 1'b1 : i_Switch_2;
  assign o_LED_3 = (r_SM_Main == PATTERN_SHOW &&
                    r_Pattern[r_Index] == 2'b10) ? 1'b1 : i_Switch_3;
  assign o_LED_4 = (r_SM_Main == PATTERN_SHOW &&
                    r_Pattern[r_Index] == 2'b11) ? 1'b1 : i_Switch_4;
--snip--
```

VHDL
```
--snip--
  o_LED_1 <= '1' when (r_SM_Main = PATTERN_SHOW and
                       r_Pattern(r_Index) = "00") else i_Switch_1;
  o_LED_2 <= '1' when (r_SM_Main = PATTERN_SHOW and
                       r_Pattern(r_Index) = "01") else i_Switch_2;
  o_LED_3 <= '1' when (r_SM_Main = PATTERN_SHOW and
                       r_Pattern(r_Index) = "10") else i_Switch_3;
  o_LED_4 <= '1' when (r_SM_Main = PATTERN_SHOW and
                       r_Pattern(r_Index) = "11") else i_Switch_4;
--snip--
```

Here we have four continuous assignment statements, one for each LED. In each one, we use the ternary operator (?) in Verilog or when/else in VHDL to illuminate the LED in one of two cases. First, we'll drive the LED high if we're in the PATTERN_SHOW state and the value at the current index of r_Pattern matches the current LED. This will only ever be true for one LED at a time, so only one LED can be illuminated during each PATTERN_SHOW. Second, if we aren't in the PATTERN_SHOW state, the LED will be driven based on the input from its associated switch. This way the LED will light up when the player presses the corresponding switch, giving them visual feedback about the pattern they're entering.

The next part of the code uses falling edge detection to identify time-outs and button presses:

Verilog
```
--snip--
  // Create registers to enable falling edge detection
  always @(posedge i_Clk)
  begin
❶ r_Toggle   <= w_Toggle;
❷ r_Switch_1 <= i_Switch_1;
    r_Switch_2 <= i_Switch_2;
    r_Switch_3 <= i_Switch_3;
    r_Switch_4 <= i_Switch_4;

❸ if (r_Switch_1 & !i_Switch_1)
      begin
        r_Button_DV <= 1'b1;
        r_Button_ID <= 0;
      end
    else if (r_Switch_2 & !i_Switch_2)
    begin
        r_Button_DV <= 1'b1;
        r_Button_ID <= 1;
    end
    else if (r_Switch_3 & !i_Switch_3)
    begin
        r_Button_DV <= 1'b1;
        r_Button_ID <= 2;
    end
    else if (r_Switch_4 & !i_Switch_4)
    begin
        r_Button_DV <= 1'b1;
        r_Button_ID <= 3;
    end
❹ else
      begin
        r_Button_DV <= 1'b0;
        r_Button_ID <= 0;
      end
  end
--snip--
```

VHDL
```
--snip--
  -- Create registers to enable falling edge detection
  process (i_Clk) is
  begin
```

```
       if rising_edge(i_Clk) then
❶ r_Toggle   <= w_Toggle;
❷ r_Switch_1 <= i_Switch_1;
   r_Switch_2 <= i_Switch_2;
   r_Switch_3 <= i_Switch_3;
   r_Switch_4 <= i_Switch_4;

❸ if r_Switch_1 = '1' and i_Switch_1 = '0' then
       r_Button_DV <= '1';
       r_Button_ID <= "00";
   elsif r_Switch_2 = '1' and i_Switch_2 = '0' then
       r_Button_DV <= '1';
       r_Button_ID <= "01";
   elsif r_Switch_3 = '1' and i_Switch_3 = '0' then
       r_Button_DV <= '1';
       r_Button_ID <= "10";
   elsif r_Switch_4 = '1' and i_Switch_4 = '0' then
       r_Button_DV <= '1';
       r_Button_ID <= "11";
❹ else
       r_Button_DV <= '0';
       r_Button_ID <= "00";
   end if;
 end if;
end process;
--snip--
```

Notice we are still using the rising edge of the clock; we're just look-ing for falling edges for our timeouts and button presses. Recall that this falling edge is used to progress through the state machine. We perform falling edge detection on the output of the Count_And_Toggle module, where the output represents the timer expiring. We do this by first registering its output, w_Toggle, and assigning it to r_Toggle ❶. (The actual instantiation of the Count_And_Toggle module will be handled momentarily.) This creates a one-clock-cycle-delayed version of w_Toggle on r_Toggle. Then, as shown previously, we look for the condition where the current value (w_Toggle) is low, but the previous value (r_Toggle) is high. We used this earlier to trigger transitions out of PATTERN_OFF and PATTERN_SHOW.

For our switches, when a switch is pressed, it has the value 1; when a switch is not pressed, it has the value 0. We are looking for the situa-tion in which the switch goes from a 1 to a 0, which is the falling edge of the switch, representing the switch being released. We register each switch ❷ in order to detect the falling edge from the switch being released. This is followed by the actual edge detection logic ❸. For each switch, when we see a falling edge, we drive r_Button_DV high. As you've seen elsewhere in the code, this signal serves as a flag to indicate that some switch, any switch, has been released. We also set r_Button_ID to the switch's 2-bit binary code, so we'll know which switch it was. The else statement ❹ clears r_Button_DV and r_Button_ID to get ready for the next falling edge.

I've chosen to make the state machine react to button releases rather than button presses. You could try inverting the test cases ❸ to see the difference. I think you'll find it a bit unnatural if the game responds the moment a button is pressed instead of the moment it's released.

The final section of the code instantiates the Count_And_Toggle and LFSR_22 modules. Remember, you've seen the code for these modules before, in Chapter 6:

Verilog

```
--snip--
    // w_Count_En is high when state machine is in
    // PATTERN_SHOW state or PATTERN_OFF state, else low
❶ assign w_Count_En = (r_SM_Main == PATTERN_SHOW ||
                        r_SM_Main == PATTERN_OFF);

❷ Count_And_Toggle #(.COUNT_LIMIT(CLKS_PER_SEC/4)) Count_Inst
    (.i_Clk(i_Clk),
     .i_Enable(w_Count_En),
     .o_Toggle(w_Toggle));

    // Generates 22-bit-wide random data
❸ LFSR_22 LFSR_Inst
    (.i_Clk(i_Clk),
     .o_LFSR_Data(w_LFSR_Data),
   ❹ .o_LFSR_Done()); // leave unconnected

endmodule
```

VHDL

```
--snip--
    -- w_Count_En is high when state machine is in
    -- PATTERN_SHOW state or PATTERN_OFF state, else low
❶ w_Count_En <= '1' when (r_SM_Main = PATTERN_SHOW or
                          r_SM_Main = PATTERN_OFF) else '0';

❷ Count_Inst : entity work.Count_And_Toggle
    generic map (
      COUNT_LIMIT => CLKS_PER_SEC/4)
    port map (
      i_Clk    => i_Clk,
      i_Enable => w_Count_En,
      o_Toggle => w_Toggle);

    -- Generates 22-bit-wide random data
❸ LFSR_Inst : entity work.LFSR_22
    port map (
      i_Clk      => i_Clk,
      o_LFSR_Data => w_LFSR_Data,
    ❹ o_LFSR_Done => open); -- leave unconnected

❺ o_Score <= std_logic_vector(r_Score);

end RTL;
```

First we instantiate the `Count_And_Toggle` module ❷. As you saw in Chapter 6, it measures out a set amount of time by incrementing a register on each clock cycle until it reaches the `COUNT_LIMIT` parameter/generic. Here we've set `COUNT_LIMIT` to `CLKS_PER_SEC/4` to make each `PATTERN_OFF` and `PATTERN_SHOW` state last a quarter of a second, but feel free to change this to make the game run faster or slower. Keep in mind that `CLKS_PER_SEC/4` is a constant (in this case, 25,000,000 / 4 = 6,250,000) that the synthesis tools will calculate in advance, so the division operation (which would require a lot of resources) won't have to be performed inside the FPGA itself. The continuous assignment of `w_Count_En` ❶ only enables the counter during the `PATTERN_OFF` and `PATTERN_SHOW` states, since we don't want it running during other phases of the game.

Next, we instantiate the `LFSR_22` module ❸. Recall from Chapter 6 that this module has two outputs: `o_LFSR_Data` for the data itself, and `o_LFSR_Done` to signal each repetition of the LFSR's cycle. For this project we don't need `o_LFSR_Done`, so we leave the unused output unconnected in Verilog, or use the `open` keyword in VHDL ❹. When we write general-purpose modules like this, we won't always need every single output in every single application. When we don't use an output, the synthesis tools are intelligent enough to remove the associated logic, so there's no hit to our resource utilization when we have unused code.

Finally, in VHDL we need to perform one more action: converting `r_Score`, which is an `unsigned` type, to a `std_logic_vector` so we can assign that value to `o_Score` ❺. Because VHDL is strongly typed, you'll frequently see type conversions like this when looking at VHDL code.

Testing the Memory Game

The code and state machine diagram that we've been looking at here represent the final version of this game, but it went through some improvements and fixes as I developed it. A lot of the changes were a result of me actually playing the game and experimenting to see what I liked and didn't like. For example, when I was first designing the state machine, I went straight from `START` to `PATTERN_SHOW` without passing through `PATTERN_OFF`. This made the first LED come on immediately, which was confusing; it was hard to tell whether the game had started or not. So, I switched the order to add a delay at the outset.

Most of the changes that I made followed this same pattern: program the board, play the game, see behavior I don't like, change the code, play the game. Another example is that the LED on-time was too long initially, so I decreased it to make the gameplay snappier. These sorts of issues are more about feel; running a simulation wouldn't have identified them.

Simulations and testbenches are valuable for understanding where and why bugs are occurring and how to fix them. Most of my "problems" weren't bugs, but behaviors that I wanted to change based on my experience playing the game. I did create a testbench that allowed me to simulate button presses, however, to see how the `State_Machine_Game` module responded. That code is available in the book's GitHub repository, if you'd like to look at it. It's a simple testbench, not one that performs any self-checking, but it did help me find a few bugs when I was initially writing this state machine.

Adding the Pin Constraints

Since we've added a new interface at the highest level (the seven-segment display), we need to add those signals to our physical constraints file. If we forget this step, the place and route tool will likely automatically choose pins for us, which will almost certainly be incorrect. You'll have to refer to the schematic for your development board to trace the signal paths from the seven-segment display back to your FPGA for each signal. Here are the constraints needed for the Go Board, for example:

```
set_io o_Segment2_A 100
set_io o_Segment2_B 99
set_io o_Segment2_C 97
set_io o_Segment2_D 95
set_io o_Segment2_E 94
set_io o_Segment2_F 8
set_io o_Segment2_G 96
```

See Chapter 2 for a reminder on how to add physical constraints to your iCEcube2 project.

Building and Programming the FPGA

At this point, we're ready to build the code for the FPGA. Let's take a look at the synthesis results of both the Verilog and the VHDL. Your report should look similar to the following:

Verilog
```
--snip--
Register bits not including I/Os: 164 (12%)
--snip--
Total LUTs: 239 (18%)
```

VHDL
```
--snip--
Register bits not including I/Os: 163 (12%)
--snip--
Total LUTs: 225 (17%)
```

The results are pretty close between the Verilog and VHDL versions; we're using about 12 percent of the available flip-flops and 18 percent of the available LUTs for this project. We have an entire memory game consisting of a few hundred lines of code, and we've used less than 20 percent of the FPGA's main resources. Not bad!

Program your development board and play the game. See if you can beat it, and if you can, try increasing the difficulty by changing GAME_LIMIT up to its maximum difficulty of 11. I found it quite challenging!

Summary

In this chapter you've learned about state machines, which are critical building blocks in many programming disciplines, including FPGAs. State machines are used to precisely control the flow through a sequence of operations. The operations are organized into a network of states, with events that trigger transitions between those states. After reviewing a simple example, you designed and implemented a sophisticated state machine to control a memory game for your development board. The project combined many elements that we've discussed throughout the book, including debounce logic to clean up the inputs from the switches, an LFSR for pseudorandom number generation, and a counter to keep track of time. You also learned to use a seven-segment display to create the game's scoreboard.

9

USEFUL FPGA PRIMITIVES

So far, you've learned about the two most fundamental FPGA components: the LUT and the flip-flop. These general-purpose components are the main workhorses in your FPGA, but there are also other dedicated components that are commonly used in FPGA designs for more specialized tasks. These components are usually called *primitives*, but they're also sometimes referred to as *hard IP* or *cores*.

Working with primitives helps you get the most out of your FPGA. In fact, a lot of modern FPGA development revolves around linking together these pre-existing primitives, with custom code added as needed for application-specific logic. In this chapter, we'll explore three important primitives: the block RAM (BRAM), the digital signal processor (DSP) block, and the phase-locked loop (PLL). You'll learn what role each one plays within an FPGA and see how to create them through your Verilog or VHDL code, or with assistance from your build tools.

The primitives we'll discuss are especially important on higher-end FPGAs, more advanced than the iCE40 FPGAs we've focused on so far. With these feature-rich FPGAs, the companion GUI software has become a critical piece of the build process. These GUIs are more complicated than the iCEcube2 tool we've been working with, and a large part of the complexity stems from the creation and wiring up of these primitives. Once you have an understanding of how the primitives work, however, you'll be better equipped to start working with these more advanced tools and to take full advantage of the common built-in features of professional-grade FPGAs.

How to Create Primitives

There are a few different ways to create an FPGA primitive. Up to this point, we've been writing Verilog or VHDL and letting the synthesis tools decide for us how to translate that code into primitives. We trust the tools to understand when we want to create a flip-flop or a LUT. This is called *inference*, since we're letting the tools infer (or make an educated guess about) what we want based on our code.

In general, the tools are able to understand our intentions quite well. However, there are some primitives that the synthesis tools won't be able to infer. To create those, you need to use another method: you can either explicitly instantiate the primitive in your code or use the GUI built into most synthesis tools to automate the creation process.

Instantiation

Instantiation is the creation of a primitive from a code template written by the FPGA manufacturer. When you instantiate a primitive component, it looks like you're instantiating a Verilog or VHDL module—but in this case, the module you're instantiating isn't one you've created. Rather, it's built into the tools for your specific FPGA. The actual module code behind these primitives is often unavailable to view; it's part of the secret sauce that the FPGA vendors like to keep to themselves.

Let's look at an example of how to instantiate a block RAM (we'll talk more about these primitives later in the chapter):

Verilog

```
RAMB18E1 #(
  // Address Collision Mode: "PERFORMANCE" or "DELAYED_WRITE"
  .RDADDR_COLLISION_HWCONFIG("DELAYED_WRITE"),
  // Collision check: Values ("ALL", "WARNING_ONLY", "GENERATE_X_ONLY" or "NONE")
  .SIM_COLLISION_CHECK("ALL"),
  // DOA_REG, DOB_REG: Optional output register (0 or 1)
  .DOA_REG(0),
  .DOB_REG(0),
  // INITP_00 to INITP_07: Initial contents of parity memory array
.INITP_00(256'h0000000000000000000000000000000000000000000000000000000000000000),
--snip--
```

```
.INIT_3F(256'h0000000000000000000000000000000000000000000000000000000000000000),
   // INIT_A, INIT_B: Initial values on output ports
   .INIT_A(18'h00000),
   .INIT_B(18'h00000),
   // Initialization File: RAM initialization file
   .INIT_FILE("NONE"),
   // RAM Mode: "SDP" or "TDP"
   .RAM_MODE("TDP"),
   // READ_WIDTH_A/B, WRITE_WIDTH_A/B: Read/write width per port
   .READ_WIDTH_A(0),              // 0-72
   .READ_WIDTH_B(0),              // 0-18
   .WRITE_WIDTH_A(0),             // 0-18
   .WRITE_WIDTH_B(0),             // 0-72
   // RSTREG_PRIORITY_A, RSTREG_PRIORITY_B: Reset or enable priority ("RSTREG" or "REGCE")
   .RSTREG_PRIORITY_A("RSTREG"),
   .RSTREG_PRIORITY_B("RSTREG"),
   // SRVAL_A, SRVAL_B: Set/reset value for output
   .SRVAL_A(18'h00000),
   .SRVAL_B(18'h00000),
   // Simulation Device: Must be set to "7SERIES" for simulation behavior
   .SIM_DEVICE("7SERIES"),
   // WriteMode: Value on output upon a write ("WRITE_FIRST", "READ_FIRST", or "NO_CHANGE")
   .WRITE_MODE_A("WRITE_FIRST"),
   .WRITE_MODE_B("WRITE_FIRST")
)
RAMB18E1_inst (
   // Port A Data: 16-bit (each) output: Port A data
   .DOADO(DOADO), ❶             // 16-bit output: A port data/LSB data
   .DOPADOP(DOPADOP),            // 2-bit output: A port parity/LSB parity
   // Port B Data: 16-bit (each) output: Port B data
   .DOBDO(DOBDO),                // 16-bit output: B port data/MSB data
   .DOPBDOP(DOPBDOP),            // 2-bit output: B port parity/MSB parity
   // Port A Address/Control Signals: 14-bit (each) input: Port A address and control signals
   // (read port when RAM_MODE="SDP")
   .ADDRARDADDR(ADDRARDADDR),    // 14-bit input: A port address/Read address
   .CLKARDCLK(CLKARDCLK),        // 1-bit input: A port clock/Read clock
--snip--
```

VHDL

```
RAMB18E1_inst : RAMB18E1
generic map (
   -- Address Collision Mode: "PERFORMANCE" or "DELAYED_WRITE"
   RDADDR_COLLISION_HWCONFIG => "DELAYED_WRITE",
   -- Collision check: Values ("ALL", "WARNING_ONLY", "GENERATE_X_ONLY" or "NONE")
   SIM_COLLISION_CHECK => "ALL",
   -- DOA_REG, DOB_REG: Optional output register (0 or 1)
   DOA_REG => 0,
   DOB_REG => 0,
   -- INITP_00 to INITP_07: Initial contents of parity memory array
   INITP_00 => X"0000000000000000000000000000000000000000000000000000000000000000",
--snip--
   INIT_3F => X"0000000000000000000000000000000000000000000000000000000000000000",
```

```
-- INIT_A, INIT_B: Initial values on output ports
INIT_A => X"00000",
INIT_B => X"00000",
-- Initialization File: RAM initialization file
INIT_FILE => "NONE",
-- RAM Mode: "SDP" or "TDP"
RAM_MODE => "TDP",
-- READ_WIDTH_A/B, WRITE_WIDTH_A/B: Read/write width per port
READ_WIDTH_A => 0,            -- 0-72
READ_WIDTH_B => 0,            -- 0-18
WRITE_WIDTH_A => 0,           -- 0-18
WRITE_WIDTH_B => 0,           -- 0-72
-- RSTREG_PRIORITY_A, RSTREG_PRIORITY_B: Reset or enable priority ("RSTREG" or "REGCE")
RSTREG_PRIORITY_A => "RSTREG",
RSTREG_PRIORITY_B => "RSTREG",
-- SRVAL_A, SRVAL_B: Set/reset value for output
SRVAL_A => X"00000",
SRVAL_B => X"00000",
-- Simulation Device: Must be set to "7SERIES" for simulation behavior
SIM_DEVICE => "7SERIES",
-- WriteMode: Value on output upon a write ("WRITE_FIRST", "READ_FIRST", or "NO_CHANGE")
WRITE_MODE_A => "WRITE_FIRST",
WRITE_MODE_B => "WRITE_FIRST"
)
port map (
  -- Port A Data: 16-bit (each) output: Port A data
  DOADO => DOADO, ❶               -- 16-bit output: A port data/LSB data
  DOPADOP => DOPADOP,            -- 2-bit output: A port parity/LSB parity
  -- Port B Data: 16-bit (each) output: Port B data
  DOBDO => DOBDO,               -- 16-bit output: B port data/MSB data
  DOPBDOP => DOPBDOP,           -- 2-bit output: B port parity/MSB parity
  -- Port A Address/Control Signals: 14-bit (each) input: Port A address and control signals
  -- (read port when RAM_MODE="SDP")
  ADDRARDADDR => ADDRARDADDR,   -- 14-bit input: A port address/Read address
  CLKARDCLK => CLKARDCLK,       -- 1-bit input: A port clock/Read clock
--snip--
```

This code is an example of instantiation of a RAMB18E1 component (a type of block RAM) from an AMD FPGA. The code makes the block RAM available for use by wiring its internal signals to signals external to the block RAM: for example, it wires the block RAM's internal DOADO signal, a 16-bit output, to an external signal of the same name ❶. I've omitted many more lines of code that make similar connections. It's not important that you understand the details of this code; it's just to demonstrate what instantiation looks like. Clearly a block RAM is a complicated component, with many bells and whistles available to you. Instantiation specifies every single input and output of the primitive and allows you to set them exactly as you want. However, it also requires that you have a deep knowledge of the primitive being instantiated. If you connect it improperly, it won't work as intended.

If you wanted to, it would be possible to instantiate, rather than infer, even a simple component like a flip-flop. Here's what AMD's Verilog template looks like for instantiating a single flip-flop (which AMD calls an FDSE):

```
FDSE #(
   .INIT(1'b0) // Initial value of register (1'b0 or 1'b1)
) FDSE_inst (
   .Q(Q),      // 1-bit data output
   .C(C),      // 1-bit clock input
   .CE(CE),    // 1-bit clock enable input
   .S(S),      // 1-bit synchronous set input
   .D(D)       // 1-bit data input
);
```

VHDL

```
FDSE_inst : FDSE
generic map (
   INIT => '0') -- Initial value of register ('0' or '1')
port map (
   Q => Q,      -- Data output
   C => C,      -- Clock input
   CE => CE,    -- Clock enable input
   S => S,      -- Synchronous set input
   D => D       -- Data input
);
```

Notice that this primitive has the normal connections we'd expect from a flip-flop, including the data output (Q), the clock input (C), the clock enable (CE), and the data input (D). After instantiating this flip-flop, you could then make use of these connections in your code. If you had to instantiate every single flip-flop in your entire FPGA, however, it would take quite a lot of code!

NOTE *I found the templates for the RAM18E1 block RAM and the FDSE flip-flop in AMD's online Libraries Guide, which contains the templates for all primitives throughout AMD FPGAs. Every FPGA manufacturer has a similar resource where you'll find the instantiation templates for its primitives.*

The benefit of instantiating a primitive is that it gives you exactly what you want. You don't need to trust the synthesis tools to guess at what you're trying to do. However, there are clearly some downsides. As you've just seen, instantiation takes more code than inference. It also requires you to wire up every connection correctly, or the design won't function as intended. This means you need to understand the primitive at a deep level. Finally, each primitive needs to be instantiated using a dedicated template specific to your FPGA vendor, or sometimes specific to just a subset of devices within a family of FPGAs. For example, the RAMB18E1 block RAM component we instantiated earlier only exists on AMD FPGAs; Intel and Lattice FPGAs have their own block RAMs. Therefore, instantiation makes your code less portable than writing more generic Verilog or VHDL where the tools can just infer the primitive based on which FPGA you're targeting. Next, we'll look at the alternative: using the GUI.

The GUI Approach

Every FPGA vendor has its own GUI or IDE for FPGA development, and that GUI will have a section allowing you to view the library of available primitives for your FPGA. You can select a primitive that you want to add to your project, and the tool will walk you through the process. Additionally, the GUI explains how the primitive works and what each setting controls. Figure 9-1 shows an example of creating a block RAM using the Lattice Diamond GUI. As mentioned in Chapter 2, this is Lattice's IDE for working with higher-end FPGAs with features like the primitives discussed in this chapter. (The iCEcube2 IDE doesn't have a GUI for creating primitives, since it's designed to work primarily with simpler FPGAs.)

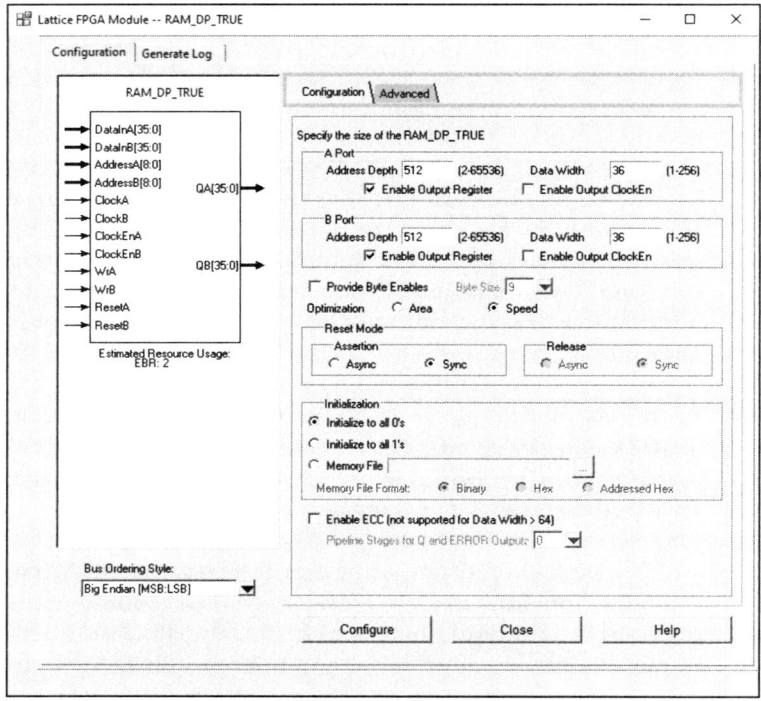

Figure 9-1: Instantiating a block RAM with a GUI

The block diagram on the left side of the window visually demonstrates the block RAM's inputs and outputs. In the configuration section on the right, it's clear which selections for the primitive are mutually exclusive. These are represented with radio buttons, like Initialize to All 0's or Initialize to All 1's. We can also tell which options can be enabled or disabled. These are represented by checkboxes, like Enable Output Register or Enable Output ClockEn. In addition, there's a convenient Help button in the bottom-right corner that can guide you through some of these decisions if you're unsure what to pick.

Once you've configured a primitive with a GUI, you'll get an instantiation template that you can drop into your Verilog or VHDL code, much like the one we looked at in the previous section. The template will be customized to the exact settings that you picked in the GUI so you can wire up your primitive without having to make any guesses about how to configure it.

Compared to direct instantiation, the GUI method is more approachable for beginners. You're much less likely to make a mistake using the GUI, since you have the menus to guide you, but you can still control exactly what you get, just like with instantiation. There is an important downside to this approach, however. If you need to change a setting in your primitive, then you need to open the GUI and run through the whole process again. This might not sound like a big deal, but if your design features many primitives created using a GUI, making adjustments can become quite tedious and time-consuming.

The Block RAM

A *block RAM (BRAM)* is a dedicated memory storage component built into your FPGA. Next to LUTs and flip-flops, block RAMs are the third most common FPGA primitive. We touched briefly on block RAMs in Chapter 6, when we discussed common memory modules like RAMs and FIFOs. As I mentioned in that chapter, when you need a memory over a certain size, it will be created using a block RAM instead of flip-flops.

Creating memory for storing data is an incredibly common task in FPGAs. You might use a block RAM for storing read-only data, like calibration values, or you might regularly write data to a block RAM from an off-chip device like an analog-to-digital converter (ADC) and then read from it later. Block RAMs are also commonly used to buffer data between a producer and a consumer, including when sending data between clock domains. In this case, the block RAM can be configured as a FIFO, with features specially designed to handle the metastability issues that arise when crossing between domains (we discussed how you can transmit data across cross clock domains back in Chapter 7).

The number of block RAMs available, and the specific features of each block RAM, will vary from FPGA to FPGA and vendor to vendor. You should always consult your FPGA's datasheet and memory guide for details particular to your model. As an example, Figure 9-2 shows a datasheet highlighting the block RAMs on Intel's Cyclone V line of FPGAs.

	Product Line	Cyclone V E FPGAs[1]		
		5CEA2	5CEA4	5CEA5
Resources	LEs (K)	25	49	77
	ALMs	9,434	18,480	29,080
	Registers	37,736	73,920	116,320
	M10K memory blocks	176	308	446
	M10K memory (Kb)	1,760	3,080	4,460
	MLAB memory (Kb)	196	303	424
	Variable-precision DSP blocks	25	66	150
	18 x 18 multipliers	50	132	300
	Global clock networks	16	16	16
	PLLs[2] (FPGA)	4	4	6

Figure 9-2: Block RAMs on the Cyclone V product line

Intel refers to block RAMs as *memory blocks.* The first of the highlighted lines in the datasheet is telling us how many of these memory blocks are available on each of three FPGA models: 176 on the 5CEA2 FPGA, 308 on the 5CEA4, and 446 on the 5CEA5 part. The next line on the datasheet shows the total number of kilobits (Kb) of block RAM storage available. Each memory block holds 10Kb (hence the *M10K* in the name), so there are 1,760Kb of BRAM storage on the 5CEA2 FPGA, 3,080Kb on the 5CEA4, and 4,460Kb on the 5CEA5.

You might be surprised by how little storage that really is. Even the largest amount, 4,460Kb, is less than a megabyte! Consider the fact that you can get a 32-gigabyte MicroSD card, which has thousands of times more storage space, for around $10, and you'll start to appreciate that FPGAs aren't designed for storing data in any significant quantity. Rather, block RAMs are there to buffer data on the FPGA for temporary usage. If you need to store large amounts of data, you'll have to use an external chip to do that. MicroSD cards, DDR memory, SRAM, and flash memory are common examples of chips that an FPGA might interface to in order to expand its memory storage and retrieval capabilities.

You should also notice in Figure 9-2 that block RAMs are the fourth item in the Cyclone V datasheet's list of FPGA resources, after LEs, ALMs, and registers. Those are the terms that Intel uses to describe LUTs and flip-flops (LE stands for logic element and ALM for Adaptive Logic Module). While you may not always need many block RAMs for your application, this prime position in the datasheet highlights that block RAMs are often one of the most significant primitive components to take into consideration when choosing an FPGA.

Features and Limitations

There are some common features and some limitations that are helpful to keep in mind when working with block RAMs. First, block RAMs usually come in only one size on an FPGA; 16Kb per block RAM is common. This one-size-fits-all approach means that if you only need to use 4Kb out of the

16Kb, you'll still use up an entire block RAM primitive. There's no way to divide a single block RAM component into multiple memories, and in that way, block RAMs can be limiting.

In other ways, however, block RAMs can be quite flexible. You can store data in whatever width you like: for example, with a 16Kb block RAM you can store data that's 1 bit wide and 16,384 bits (2^{14}) deep, or 8 bits wide and 2,048 deep, or 32 bits wide and 512 deep, among other possibilities. It's also possible to create memories that are larger than a single block RAM. For example, if you needed to store 16 kilobytes (KB) of data, that would use up eight individual block RAMs (16Kb × 8 = 16KB). The tools are smart enough to cascade the block RAMs and make them look like one large memory, rather than eight individual components that you need to index into individually.

Other common features include error detection and correction, where the block RAM has some extra bits reserved to detect and correct any errors that might occur within the memory itself (that is, when a 1 changes to a 0, or vice versa). If that happens in your memory, a value could be completely corrupted and produce very strange behavior when the FPGA tries to analyze it.

Error detection and correction are two separate but related processes: the FPGA can *detect* some number of bit errors and notify you about their presence, and, separately, it can automatically *correct* some number of bit errors. The number of bit errors that can be corrected is usually less than the number of bit errors that can be detected. The important thing here is that error detection and correction within a block RAM are performed automatically, without you having to do anything.

Many block RAMs can also be initialized to default values. This can be a useful feature if you need to store a large number of initial values or if you want to create read-only memory (ROM). Pushing those values to a block RAM rather than taking up flip-flops for data storage can be a valuable way to save resources. We touched on this idea back in Chapter 7, when we were looking at parts of Verilog and VHDL that are synthesizable and not synthesizable. Even though reading from a file is normally not synthesizable—remember that there's no filesystem on an FPGA unless you create it yourself—we can read data from files as part of the synthesis process to preload a block RAM with default values. Again, I recommend consulting the memory guide for your particular FPGA to find out which features it supports.

Creation

When using a block RAM in your design, I generally recommend inferring it. As you saw in Chapter 6, when we create a two-dimensional memory element, the tools will easily recognize it. Whether or not this memory gets pushed to a block RAM depends on its size. Again, the synthesis tool is smart about this: it knows how many bits of memory you're creating, and if it's above some threshold, then it will be pushed to a block RAM.

Otherwise, the tool will just use flip-flops. For example, if you're creating a memory that holds 16 bytes, it will likely be pushed to flip-flops. You only need $16 \times 8 = 128$ bits of memory, so it doesn't make much sense to use an entire 16Kb block RAM for this small quantity of data.

At which point the tools will start pushing memory to block RAMs instead of using flip-flops is highly dependent on the situation. To find out what your tool decided for a particular design, consult your utilization report after synthesis. Here's an example:

```
--snip--
Number of registers:   1204 out of 84255 (1%)
--snip--
Number of LUT4s:       1925 out of 83640 (2%)
--snip--
❶ Number of block RAMs: 3 out of 208 (1%)
```

The utilization report lists the number of block RAMs required ❶, just as it lists the number of flip-flops (registers) and LUTs (LUT4s, or four-input LUTs in this case). If you see that no block RAMs are being used, then your memory was inferred as flip-flops instead. As a reminder, I always recommend double-checking your utilization report to make sure the tools are inferring what you expect.

If you're wary of trying to infer large memory elements, or you're confused about which features you may or may not want to take advantage of in your block RAM, then creating it with a GUI is your best option. The GUI will guide you through the process, so for beginners it's very helpful. Using the GUI is also the best way to ensure that you're using a FIFO correctly when crossing clock domains, as it can help you handle the complexities involved.

The Digital Signal Processing Block

Digital signal processing (DSP) is a catch-all term for performing math-based operations on signals within a digital system. Often these math operations need to happen very fast and in parallel, which makes FPGAs an excellent tool for the job. Since DSP is such a common FPGA application, another kind of FPGA primitive, the *DSP block*, exists for this purpose. DSP blocks (also known as *DSP tiles*) specialize in performing mathematical operations, in particular *multiply–accumulate (MAC)*, which is an operation where a multiplication is followed by an addition. Before we look more closely at these primitives, however, it's worth taking a step back to discuss the difference between analog and digital signals.

Analog vs. Digital Signals

An *analog signal* is a continuous signal representing some physical measurement. A common example is the audio signal stored on a vinyl record (a big black shiny thing that has music on it, sometimes seen in old movies or new hipster bars). The record is etched with a continuous groove that mirrors the continuous waveform of the audio. Then a record player reads that waveform with a needle and amplifies the resulting signal to play back the sound. The information is always analog; no conversion is needed.

Digital signals, on the other hand, aren't continuous. Rather, they consist of discrete measurements at individual points in time, with gaps in between. A common example is the audio signal stored on a CD, where the sound is represented as a series of 1s and 0s. If you have enough discrete measurements, you can fill in the gaps to create a reasonably accurate approximation of an analog signal from those digital values. A CD player reads those digital values and rebuilds an analog waveform from them. The result is always an approximation of the original analog signal, however, which is why some audiophiles prefer the true analog signal of a record to digital media like CDs and MP3s.

Within your common FPGA fabric, like LUTs and flip-flops, data is represented digitally. So what do you do if you have some analog signal that you need to bring into your FPGA? This is the purpose of an ADC: it converts an analog signal into a digital one by *sampling* it, or recording its value, at discrete points in time. Figure 9-3 shows how this works.

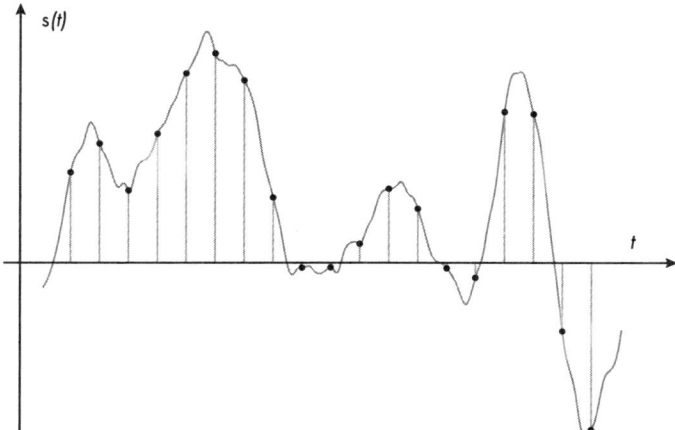

Figure 9-3: Digital sampling of an analog signal

The undulating line moving from left to right in the figure represents a continuous analog signal, and the dark points along that line represent the individual samples taken of that signal to convert it into a digital form. Notice that the samples are taken at regular time intervals. The frequency at which the analog signal is sampled is called the *sampling frequency* or *sampling rate*. The higher the sampling rate, the more accurately we can represent an analog signal, because it's easier to connect the discrete dots into something that looks like the original waveform. However, a higher sampling rate also means that we have more data that we have to process: each dot represents some number of bits of digital data, so the more dots you have, the more bits you're working with.

Common DSP Tasks

FPGAs commonly take an analog signal as input, digitize it, and then do some math to process that digital data. As an example, let's say we have an audio signal that we've sampled within our FPGA. Let's furthermore assume that the recorded data was too quiet, so when it's played back it's hard to hear. How can we manipulate the digital signal such that the output volume is louder? One simple thing we can do is multiply every digital value by some constant, say 1.6. This is called applying *gain* to a signal. How would we accomplish this within an FPGA? It's quite simple:

```
gain_adjusted <= input_signal * 1.6;
```

We take the input_signal, multiply every discrete digital value in that signal by 1.6, and store the result in the gain_adjusted output. Here is where the DSP primitive comes into play. When we write code like this, the synthesis tools will see that we're performing a multiplication operation and infer a DSP block for us automatically.

Applying gain to an input signal doesn't require parallel processing. There's only one multiplication operation per data sample, and the data samples can be processed one after the other. Often, however, you'll need to perform many mathematical operations in parallel by running several DSP blocks simultaneously. A common example is creating a *filter*, a system that performs mathematical operations on a signal to reduce or enhance certain features of the input signal. A *low-pass filter (LPF)*, for instance, keeps the frequency components of a signal that are below some cutoff while reducing the frequencies that are above that cutoff, which can be useful for removing high-frequency noise from an input signal. Lowering the treble slider on your audio system is a real-world example of applying a low-pass filter, since it will reduce high frequencies within the audio. The details of implementing a digital LPF are beyond the scope of this book, but since it requires many multiplication and addition operations all occurring at the same time, FPGAs are well suited for the task.

Another example of parallel math that might be performed in an FPGA is processing video data to create a blur effect. Blurring video involves replacing individual pixel values with the average value of a group of neighboring pixels. This requires performing math on the many pixels in an image at the same time, and this must happen quickly since the video data consists of many images per second. An FPGA is very capable of performing these parallel mathematical operations using DSP blocks.

Features

DSP blocks are versatile primitives, providing many features that facilitate different math operations. You won't always need every feature for your application—most often, you'll just be performing a multiplication or addition operation—but for more complicated scenarios, the DSP block can be set up to solve a wide range of problems. Figure 9-4 provides a detailed look at a DSP block in an FPGA from AMD. Each manufacturer's DSP primitive is a bit different, but this example is representative of the typical features available.

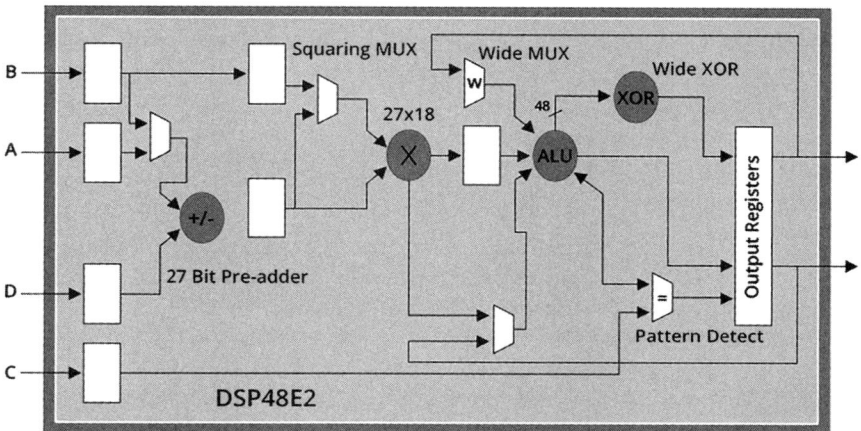

Figure 9-4: Block diagram of a DSP primitive

This diagram actually shows a simplified version of the DSP block. It's not critical to understand the complete anatomy of the primitive, but it's worth pointing out a few things. First, notice that this DSP block can take up to four inputs and has two outputs. This allows for more applications than simply multiplying two numbers together: for example, MAC, where the result of a multiplication is fed back into the input at the next clock cycle for an addition operation.

Toward the left-hand side of the block diagram, you can see a *pre-adder* block. This can be enabled if an addition operation is requested prior to another mathematical operation. To the right of this, near the middle of the diagram, is a circle with an X in it. This is the *multiplier*, which is the heart of the DSP block. It performs multiplication operations at very high speeds. To its right is a circle labeled ALU, short for *arithmetic logic unit*, which can perform more operations, like addition and subtraction. Finally, there are built-in output registers that can be enabled to sample the outputs and help meet timing at fast data rates.

Like the number of block RAMs, the number of DSP blocks available to you will vary from FPGA to FPGA. Some higher-end FPGAs have thousands of DSP blocks inside them; again, you should consult your FPGA's datasheet for details specific to your model. As an example, Figure 9-5 highlights the information on DSP blocks in the datasheet for Intel's Cyclone V product line.

	Product Line	Cyclone V E FPGAs[1]		
		5CEA2	5CEA4	5CEA5
Resources	LEs (K)	25	49	77
	ALMs	9,434	18,480	29,080
	Registers	37,736	73,920	116,320
	M10K memory blocks	176	308	446
	M10K memory (Kb)	1,760	3,080	4,460
	MLAB memory (Kb)	196	303	424
	Variable-precision DSP blocks	25	66	150
	18 x 18 multipliers	50	132	300
	Global clock networks	16	16	16
	PLLs[2] (FPGA)	4	4	6

Figure 9-5: DSP blocks on Cyclone V FPGAs

Notice that the DSP block information comes just below the block RAM information, again pointing to the importance of these primitives in FPGA development. The 5CEA2 FPGA has 25 DSP blocks, but that increases to 66 for the 5CEA4 and 150 for the 5CEA5. Each DSP block has two multipliers, so on the second highlighted line we can see that there are twice as many 18×18 multipliers (where 18 is the width of the inputs) as there are DSP blocks.

NOTE   *If there aren't any DSP blocks available on your FPGA, that doesn't mean you can't perform these types of operations. Multiplication and addition operations will just be implemented with LUTs, rather than with dedicated DSP blocks. We'll discuss this further in Chapter 10.*

Creation

As with block RAMs, I generally recommend using inference to create DSP blocks. Most of the multiplication operations you'll need to do will require two inputs and one output, as you saw earlier when we applied gain to a signal. It's simple enough to write the relevant code in Verilog or VHDL and let the tools handle the rest. Remember to check your synthesis report to ensure that you're getting what you expect, but I've had good luck with the synthesis tools understanding my intent with addition and multiplication and pushing those operations to DSPs where relevant. The user guides for your particular FPGA will also provide you with suggestions on how to write your Verilog or VHDL code to help ensure the tools understand your intentions.

If you have more complicated needs for your DSP blocks, or if you want to explore all of the features and capabilities internal to them, then you should probably create them using a GUI to ensure you get what you want. Figure 9-6 shows an example of creating a multiplier within the Lattice Diamond GUI.

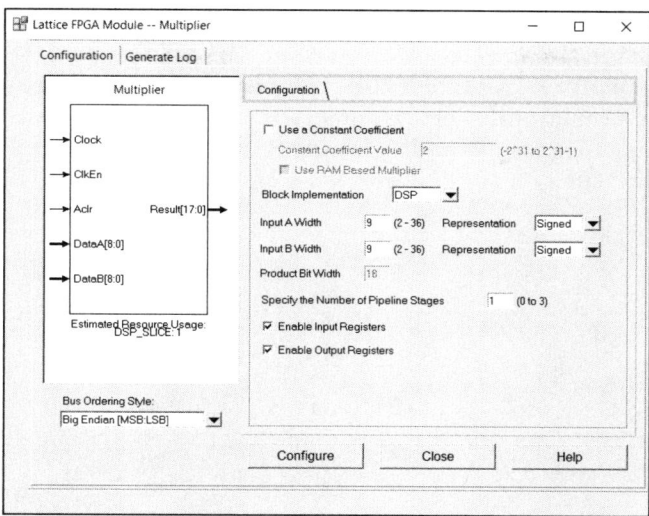

Figure 9-6: Creating a DSP block with a GUI

One thing to highlight here is the Block Implementation drop-down menu. You can change this from DSP to LUT to use look-up tables rather than a DSP block to perform this multiplication operation. As mentioned previously, LUTs and DSPs are both capable of performing math operations, including multiplication. With the DSP block, however, you'll save LUT resources, and you'll be able to run the math operation at much faster clock rates, since you'll be using a dedicated primitive highly optimized for math.

The Phase-Locked Loop

The *phase-locked loop (PLL)* is a primitive commonly used as the main clock generator for your entire FPGA. Very often, you'll have an external clock chip that runs at some frequency. On some FPGAs, you can simply use that input clock to feed all of your synchronous logic directly, as we've done in this book's projects. In this case, your logic frequency will be fixed at the frequency of whatever external clock you picked. But what happens if you need to change that frequency? Without a PLL, you would need to physically remove the external clock chip and replace it with a different component that generates the clock frequency you want to switch to. With a PLL, however, you can generate a different clock frequency inside your FPGA by changing a few lines of code, without requiring a new external component.

PLLs also make it easy to have multiple clock domains in your FPGA design. Say you have some external memory that runs at 100 MHz, but you want your main logic to run at 25 MHz. You *could* purchase a second external clock and feed that into your FPGA, but a better solution is to use a PLL, since this primitive can generate multiple clock frequencies simultaneously.

Not all FPGAs have a PLL, but many have at least one, and some have several. The datasheet will tell you what's available. As an example, Figure 9-7 highlights the PLLs available on Intel's Cyclone V product line.

Product Line		Cyclone V E FPGAs[1]		
		5CEA2	5CEA4	5CEA5
Resources	LEs (K)	25	49	77
	ALMs	9,434	18,480	29,080
	Registers	37,736	73,920	116,320
	M10K memory blocks	176	308	446
	M10K memory (Kb)	1,760	3,080	4,460
	MLAB memory (Kb)	196	303	424
	Variable-precision DSP blocks	25	66	150
	18 x 18 multipliers	50	132	300
	Global clock networks	16	16	16
	PLLs[2] (FPGA)	4	4	6

Figure 9-7: PLLs on the Cyclone V product line

The 5CEA2 and 5CEA4 FPGAs both have four PLLs, while the 5CEA5 has six. Given that each PLL can generate multiple clocks, that should be more than enough for all your clocking needs.

How It Works

A PLL serves as the source of your clock distribution throughout your FPGA by taking a single clock input, often called the *reference clock*, and generating one or more clock outputs from it. The input clock comes from a dedicated external component, and the outputs can run at completely different frequencies from the input clock and from one another. The block diagram in Figure 9-8 shows the most common signals on a PLL.

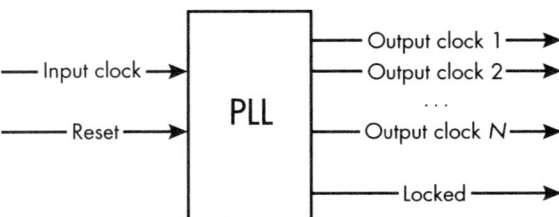

Figure 9-8: Common PLL signals

The PLL typically takes two inputs: a clock signal and a reset. The reset input will stop the PLL from running when it's asserted.

On the output side, the PLL has some number of output clocks in the range 1 to *N*, with the maximum number depending on the FPGA. The output clocks can be of different frequencies, depending on what you need for your design. These frequencies are achieved by taking the input reference clock and multiplying and/or dividing it to get the desired value. For example, say you have a 10 MHz input reference clock, and you want a 15 MHz output clock. The PLL would multiply the reference clock by 3 (giving you 30 MHz), then divide it by 2 to get down to 15 MHz. The multiplication and division terms must be integers, so it's important to realize that you can't get any arbitrary frequency out of the PLL. It isn't possible to get a π MHz clock output from a 10 MHz clock input, for example, since π is an irrational number that can't be expressed as the ratio of two integers.

Besides varying the frequency of the output clock(s), a PLL can also vary their phase. A signal's *phase* is its current position along the repeating waveform of the signal, measured as an angle from 0 to 360 degrees. It's easiest to picture what this means by comparing two signals that share a frequency but aren't aligned in time. Figure 9-9 demonstrates some common phase shifts of a clock signal.

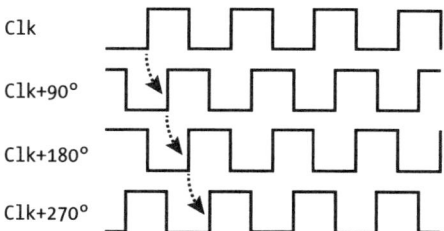

Figure 9-9: Common phase shifts

As this figure shows, shifting the phase of a clock signal results in moving the location of its rising edges. Compare the first rising edge of Clk (which has no phase shift) with the first rising edge of Clk+90° (which is phase-shifted by 90 degrees). The rising edge of Clk+90° is delayed by one-quarter of a clock period relative to Clk. Each increment of 90 degrees shifts the signal by another quarter period. Continuing the example in the figure, we have Clk+180°, which is delayed by 90 degrees from Clk+90° and 180 degrees from Clk. Notice that Clk+180° is actually the same waveform that you would get if you took the Clk signal and inverted it by swapping the highs and lows. Finally, Clk+270° is delayed by three-quarters of a clock period relative to the original Clk signal. If you went a full 360 degrees, you'd be back to your original signal. This example has demonstrated positive phase shifts, but phase can also be negative, meaning the signal is shifted backward in time compared to the other. Of course, you can shift the phase by any arbitrary angle, not just in 90-degree steps.

Creating clocks with phase shifts isn't very common in simple designs, but it can be useful in some applications. For example, it might be important for interfacing to external components, like some off-FPGA memory.

Returning to the block diagram in Figure 9-8, a PLL also typically has a *locked* output signal, which tells any module downstream that the PLL is operating and you can "trust" the clocks. It's a common design practice to use this locked signal as a reset to other modules relying on the PLL's clocks. When the PLL isn't locked, the modules downstream of the PLL are held in a reset state until the PLL is locked and ready, meaning the output clocks can be used by other modules in your FPGA design. When the PLL's reset input is driven high its locked output will go low, putting the downstream modules back into a reset condition.

If you're going to use a PLL in your design, it's a good idea to use *only* the PLL's outputs for all your clocking needs. Even if part of your design runs at the same clock frequency as the external reference clock, you shouldn't use the external clock directly to drive that part of the design. Instead, have the PLL output a clock signal at the same frequency as the external reference. By only using the PLL's outputs, you can tightly control the relationships between the output clocks. Additionally, you can confidently use the locked output of the PLL for your reset circuitry, knowing it reflects the state that *all* clocks in your design are operational.

Creation

The PLL is one primitive that I recommend using a GUI to create, since the synthesis tools won't be able to infer a PLL. Instantiation is also possible, but it's prone to errors. You need to choose PLL settings that are compatible with one another for the PLL to work successfully, and during instantiation it's easy to pick settings that won't work. If you had a 10 MHz reference clock and you wanted to generate one 15 MHz output and a separate 89 MHz output, for example, that simply might not be possible, but you might miss that fact during instantiation.

When you create a PLL using the GUI, you tell it your input reference clock and desired output clock frequencies, and the tool will tell you if it can find a solution that works. Continuing the 10/15/89 MHz example, the GUI might tell you that the closest value to 89 MHz that it can give you is 90 MHz (since 90 MHz is a multiple of both 10 MHz and 15 MHz, this is likely to work). Then it's up to you to decide whether 90 MHz will work for your design or if you really need 89 MHz, in which case you might need to use a separate PLL or change your reference clock. Figure 9-10 shows an example of the PLL GUI within Lattice Diamond.

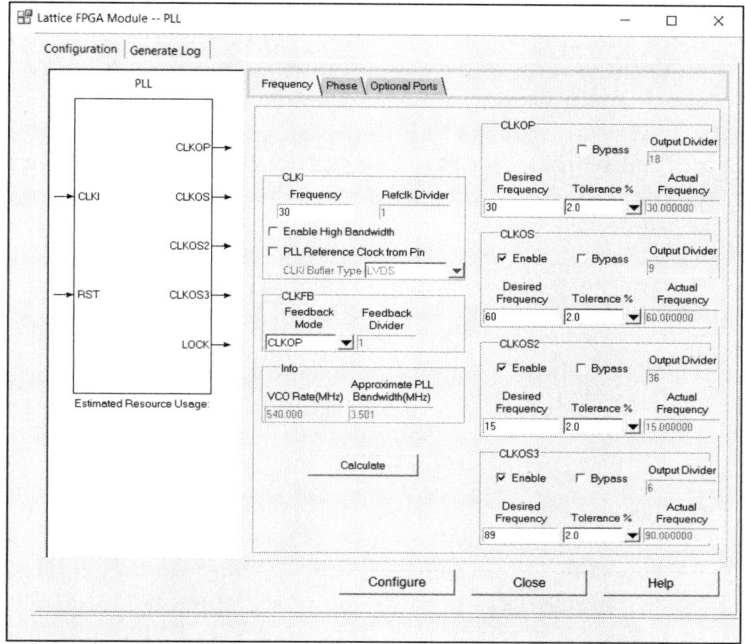

Figure 9-10: Creating a PLL with a GUI

As you can see, the GUI helps guide us through the PLL creation process. In this case, we have a 30 MHz reference on `CLKI`, and we're setting the desired output frequencies to 30 MHz on `CLKOP`, 60 MHz on `CLKOS`, 15 MHz on `CLKOS2`, and 89 MHz on `CLKOS3`. Notice that for each clock except `CLKOS3`, the actual frequency on the far right matches the desired frequency. For `CLKOS3`, when I first tried to create an 89 MHz clock with 0.0 percent tolerance, I got the error message shown in Figure 9-11.

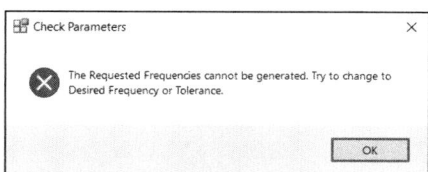

Figure 9-11: Actionable feedback from invalid PLL settings

Not until I changed the tolerance to 2.0 percent did the error message go away; the tool had selected an actual frequency of 90 MHz, which is within 2.0 percent of the requested frequency. This type of guidance isn't provided if you try to instantiate your PLL directly.

Another helpful feature of the GUI is the PLL block diagram, shown in the left half of Figure 9-10. This diagram will be updated if you modify the inputs or outputs. For example, if we disabled `CLKOS3`, that output would disappear from the block diagram to reflect that we only want to output three clock signals. This is useful to ensure you're creating what you expect. Notice that there's also a separate Phase tab near the top of the window, which allows us to specify phase shifts on our output clocks.

After designing a PLL in the GUI, you can run your design through the normal synthesis process. The utilization report will confirm you're getting a PLL, as it's one of the main primitives highlighted in the report. Here's an example:

```
--snip--
Number of PLLs:  1 out of 4 (25%)
```

This indicates one PLL is being used out of four available on this particular FPGA.

Summary

The majority of your Verilog and VHDL code will be dedicated to creating LUTs and flip-flops, which are the two most fundamental FPGA components. However, as you've seen in this chapter, FPGAs also contain other primitive components, such as block RAMs, DSP blocks, and PLLs, that add specialized functionality. Block RAMs add dedicated memory, DSP blocks enable high-speed parallel math operations, and PLLs allow you to generate different internal clock frequencies. With a combination of these FPGA building blocks, you'll be able to solve a wide range of problems efficiently.

10

NUMBERS AND MATH

Throughout this book, I've been saying that FPGAs are good at performing mathematical computations quickly. I've also been saying that FPGAs are good at doing tasks in parallel, and that the combination of these two assets—fast math done in parallel—is one of their killer features. In low-level Verilog or VHDL code, however, working with numbers and math is full of pitfalls.

In this chapter, we'll explore exactly how FPGAs manage mathematical calculations so you can avoid those pitfalls. To understand the details of how operations like addition, subtraction, multiplication, and division work, we also need to understand how to represent numbers, both positive and negative, with or without decimals, inside your FPGA. It's time for a journey in the wonderful world of computer arithmetic.

Numerical Data Types

There are many ways to represent numbers in Verilog or VHDL, as is true with all programming languages. For example, if you want to store only whole numbers, you can use an integer data type, but if you need to store fractional numbers, you'll need a data type that can represent decimals. Choosing the right type for the data you're trying to represent is critical in any programming language. If you assign data to the wrong type, you'll either get compiler errors or, worse, a design that behaves strangely. For example, trying to assign a decimal number to an integer data type could truncate the fractional component, causing an unintended rounding operation.

Additionally, you don't want to use more resources than necessary. For example, you could create every signal with a 64-bit-wide data type, but that's clearly overkill if all you need is a counter that goes from 0 to 7. FPGAs provide even more granular control over data types than you get with most other programming languages. For example, C has the `uint8_t`, `uint16_t`, and `uint32_t` data types, which create data widths of 8, 16, and 32 bits, respectively, but there's nothing in between. In Verilog and VHDL, by contrast, you can create a signal that's 9 bits wide, 15 bits wide, 23 bits wide, or any other number. We'll explore recommendations for sizing signals later in this chapter.

Representing Signed vs. Unsigned Values

When you're working with numbers, you need to know if they're positive or negative. Sometimes, such as if you're counting clock cycles to keep track of time, you'll know the values will all be positive. In this case, you can store the numbers using an *unsigned* data type; the sign (positive or negative) isn't specified and is assumed to be positive. Other times, you'll need to work with negative numbers: for example, when you're reading temperature values, where the sign of the numbers might vary. In these cases you'll need to use a *signed* data type, where the positive or negative sign is specified.

By default, signals in Verilog and VHDL are unsigned. For example, if we need a counter that counts from 0 to 7, we can declare a signal like `reg [2:0] counter;` in Verilog or `signal counter : std_logic_vector(2 downto 0);` in VHDL. We've used code like this throughout the book. It will create a 3-bit register, and since it's unsigned by default, the values on the register will all be interpreted as positive. If we want `counter` to represent negative numbers as well as positive ones, we'd have to explicitly declare it to be signed using the `signed` keyword. In Verilog we would write `reg signed [2:0] counter;`, and in VHDL we would use `signal counter : signed(2 downto 0);`.

NOTE *To access the `signed` keyword in VHDL, you need to use the `numeric_std` package, which you can do by adding the line `use ieee.numeric_std.all;` at the top of your file. You may see some code that uses the `std_logic_arith` package instead, but this isn't an official IEEE-supported library and I don't recommend using it. It's easier to make mistakes with this package than with `numeric_std`.*

Using the `signed` keyword explicitly tells the tools that this 3-bit-wide register can represent negative and positive values. But which values can we actually represent with it? Table 10-1 compares the values represented by a 3-bit unsigned register and a 3-bit signed register. (We'll discuss how to determine the signed values in the next section.)

Table 10-1: 3-Bit Unsigned vs. Signed Decimal Values

Bits	Unsigned decimal value	Signed decimal value
000	0	0
001	1	1
010	2	2
011	3	3
100	4	−4
101	5	−3
110	6	−2
111	7	−1

Notice that when a register is declared as signed, we lose some numbers on the positive end of the range (4, 5, 6, and 7 in this case) but gain some numbers on the negative end (−1, −2, −3, and −4). The range of numbers that can be represented by an unsigned register is 0 to $2^N - 1$, where N is the number of bits available. For this 3-bit register, we can represent from 0 to $2^3 - 1 = 7$ if the register is unsigned. On the other hand, the range of numbers that can be represented by a signed register is $-2^{(N-1)}$ to $2^{(N-1)} - 1$. In this case, that gives us $-2^{(3-1)}$ to $2^{(3-1)} - 1$, or $= -4$ to 3. The data is still 3 bits of binary data, but *what that binary data represents* is different.

Another feature to notice in Table 10-1 is that the values that are negative all have a 1 in the most significant bit position. In fact, the most significant bit in a signed number is the *sign bit*, which indicates whether the number being represented is positive or negative. For signed binary numbers, a 0 in the sign bit tells you that the number is positive, while a 1 in the sign bit tells you that the number is negative.

Taking the Two's Complement

How do you know what decimal value a negatively signed binary number is supposed to represent? You take its *two's complement*, a mathematical operation where you invert the number's bits and then add 1. For example, take the binary number 101. If this were an unsigned number, we'd interpret it as 5 in decimal, but if it's a signed number, the 1 in the sign bit tells us that the represented value should be negative, so we have to take the two's complement. First, we invert 101, which gives us 010. Then we add 1, giving us 011, which is 3 in decimal. Finally, we apply the negative sign to get −3. Look back at Table 10-1 and you'll see that's what we have in the row for 101.

*An alternative to the invert-and-add-one method is to start at the right-most (least significant) bit, move left until you get to the first 1, then invert all the remaining bits to the left of that 1. For example, 100010**100** becomes 011101**100**. The three bolded bits, up to and including the first 1 from the right, remain the same, while the others are inverted. In decimal, 011101100 is 236; applying the negative sign, we know that 100010100 represents −236. This method avoids the need for addition and can be simpler for long numbers.*

We can also take a two's complement to go in the other direction, converting a negative decimal number into its signed binary representation. For example, how would we represent −1 in binary using 3 bits? First, strip away the negative sign to get 1, which is 001 in binary. Then invert the bits to get 110, and add 1 to get 111. Again, check Table 10-1 and you'll see that this is the correct result.

Taking the two's complement is a useful trick us humans can use to better understand how to interpret signed numbers, but this invert-and-add-one logic *isn't something an FPGA actually does* when working with negative values. The data is binary 1s and 0s whether a number is signed or unsigned. It's just the *representation* of those 1s and 0s that makes a difference. When you have a 3-bit unsigned signal set to 101, that represents the decimal value 5. When you have a 3-bit signed signal set to 101, that represents the decimal value −3. The FPGA doesn't have to invert and add bits anywhere to know that. It just needs to know that the signal is of a signed data type. This is an important point and will become clearer as we explore mathematical operations in binary.

Sizing Signals Appropriately

When you write Verilog or VHDL code working with signed and unsigned data types, you must ensure that you properly size the signals you're creating. If you try to store too large of a number in too small of a signal, you'll get data loss. As we just discussed, for example, the value of a 3-bit unsigned counter maxes out at 7. If you try to increment it again from 7, it won't go to 8; it'll actually go back to 0. This is sometimes called *wraparound*, and if you're not expecting it, you'll end up losing count. As you'll see later in the chapter, ensuring your signals are large enough to handle your data is particularly important when the signals are for holding the results of mathematical operations.

To avoid data loss you might be tempted to make all your signals larger than they need to be, but there's a downside to this, too: you'll end up using more of your FPGA's precious resources than are needed. This may be less of a problem than you think, though. If the synthesis tools are smart enough, they might detect that your possible range of values is smaller than the signal you've created and remove the upper bits that are unused to conserve resources. If the tools did this to our counter register, for example, we'd see a warning saying something like Pruning register counter in the synthesis report. Getting a warning like this isn't usually a problem, but it might indicate code you could revisit and size differently.

As a rule of thumb, you should aim to size your signals to the values you expect them to store, but know that making them too large is a much better solution than making them too small. Of course, you have to remember that the maximum value you can represent with a given number of bits varies depending on whether the values are signed or unsigned. For comparison, Table 10-2 summarizes the ranges of possible unsigned and signed values you can represent using between 2 and 8 bits.

Table 10-2: N-bit Sizing for Unsigned and Signed Data Types

Width	Type	Min integer	Min binary	Max integer	Max binary
2	Unsigned	0	00	3	11
2	Signed	–2	10	1	01
3	Unsigned	0	000	7	111
3	Signed	–4	100	3	011
4	Unsigned	0	0000	15	1111
4	Signed	–8	1000	7	0111
5	Unsigned	0	00000	31	11111
5	Signed	–16	10000	15	01111
6	Unsigned	0	000000	63	111111
6	Signed	–32	100000	31	011111
7	Unsigned	0	0000000	127	1111111
7	Signed	–64	1000000	63	0111111
8	Unsigned	0	00000000	255	11111111
8	Signed	–128	10000000	127	01111111

Starting at a width of 2 bits, we can represent the numbers 0 to 3 unsigned, or –2 to 1 signed. At a width of 8 bits, we can represent 0 to 255 unsigned, or –128 to 127 signed.

One way to bypass the sizing dilemma is to size your signals dynamically, instead of setting them to a fixed width. We've seen some examples of this throughout the book. For instance, if you need to index into something 32 words deep but that depth could change in the future, you could write something like reg [$clog2(DEPTH)-1:0] index; instead of reg [4:0] index; in Verilog, or signal index : integer range 0 to DEPTH-1; instead of signal index : std_logic_vector(4 downto 0); in VHDL. Here, DEPTH is a parameter/generic that can be changed on the fly. Using it will generate a signal of the exact bit width you need, wide enough to index into all possible values from 0 to DEPTH-1, with no extra headroom. In this case, you would set DEPTH to 32, but if your indexing requirement were to grow to some larger value (say 1,024), the code won't break; all you'll have to do is change DEPTH. By contrast, if you arbitrarily say that index will be fixed to 8 bits wide (which has a maximum value of 255, as you can see in Table 10-2), then your code might break in the future if your requirements grow beyond that range.

Converting Between Types in VHDL

VHDL has many numeric data types, including signed and unsigned, where binary values are interpreted as positive or negative decimal numbers; integer, where you can type numbers directly into the code; and std_logic_vector, where by default binary values aren't interpreted as anything other than binary values. Because VHDL is strongly typed, you'll often need to convert between these different data types when you're working with numbers. Before we do any math, let's look at some examples of how to implement common VHDL type conversions using the numeric_std package (not the unofficial std_logic_arith).

NOTE *Verilog users don't need to worry about performing these conversions, since Verilog is loosely typed. VHDL users should consult this section as needed for reference.*

From Unsigned or Signed to Integer

This example illustrates how to convert from the unsigned or signed type to the integer type. For simplicity, we're assuming the signals are all 4 bits wide, but the conversion will work for any bit width:

```
signal in1  : unsigned(3 downto 0);
signal in2  : signed(3 downto 0);
signal out1 : integer;
signal out2 : integer;

out1 <= to_integer(in1);
out2 <= to_integer(in2);
```

For these conversions, all we need to do is call the to_integer() function from the numeric_std package. We already know the width and the sign of the input, so the output will be sized automatically. This works whether the input is unsigned (as with in1) or signed (as with in2).

From Integer to Unsigned, Signed, or std_logic_vector

This example shows how to convert from the integer type to one of the other types. Again, we're assuming 4-bit signals:

```
signal in1  : integer;
signal out1 : unsigned(3 downto 0);
signal out2 : signed(3 downto 0);
signal out3 : std_logic_vector(3 downto 0);
signal out4 : std_logic_vector(3 downto 0);

❶ out1 <= to_unsigned(in1, out1'length);
❷ out2 <= to_signed(in1, out2'length);
   -- Positive integers:
❸ out3 <= std_logic_vector(to_unsigned(in1, out3'length));
   -- Negative integers:
❹ out4 <= std_logic_vector(to_signed(in1, out4'length));
```

Here we use the to_unsigned() ❶ and to_signed() ❷ functions from numeric_std to convert from the integer to the unsigned or signed type. In addition to the value to be converted, these functions require the width of the output signal as an argument. Rather than entering the width manually, we get it by applying 'length, a VHDL attribute, to the output signal. This keeps the code flexible; if the width changes, the conversion code doesn't have to.

To get to std_logic_vector, we have to convert from integer to unsigned if the integer is positive ❸, or to signed if the integer is negative ❹. Then, once we have an unsigned or signed value of the proper width, we cast it using std_logic_vector().

From std_logic_vector to Unsigned, Signed, or Integer

Finally, here's how to convert from the std_logic_vector type to one of the other numeric types:

```
  signal in1  : std_logic_vector(3 downto 0);
  signal out1 : unsigned(3 downto 0);
  signal out2 : signed(3 downto 0);
  signal out3 : integer;
  signal out4 : integer;

❶ out1 <= unsigned(in1);
❷ out2 <= signed(in1);
  -- Demonstrates the unsigned case:
❸ out3 <= to_integer(unsigned(in1));
  -- Demonstrates the signed case:
❹ out4 <= to_integer(signed(in1));
```

To get to unsigned ❶ or signed ❷, we use a simple cast. However, VHDL needs to know if the std_logic_vector is unsigned or signed before converting to the integer type. We perform the appropriate cast using unsigned() ❸ or signed() ❹, then call the to_integer() function to do the final conversion.

Performing Mathematical Operations

Now we'll consider how basic addition, subtraction, multiplication, and division operations are performed within an FPGA, and how to implement them in Verilog and VHDL. I'll suggest some rules that, if followed, will help you avoid many of the commit pitfalls when working with binary math. The best way to explore these concepts is through examples. To that end, we'll create a large testbench that you can run in a simulator tool like EDA Playground. The testbench will execute dozens of different math equations, illustrating how binary math operations should be carried out and how they can go awry.

In general, when working with numbers and manipulating them with algebraic operations, testbenches are a very powerful tool. Hidden math issues in your code can manifest themselves in strange ways. Testbenches allow you to stress your design by running through a large range of possible inputs, to see how the code works. I find it valuable to inject data into my test-benches that stress the math operations over a wide range of values, including minimum and maximum inputs. This helps ensure a design is robust.

Before we do any math, let's set up our testbench, called Math_Examples, by declaring all the necessary inputs and outputs, as well as some helper functions in the VHDL version. This setup code provides the skeleton for the examples that follow throughout the rest of the chapter. The code for each example will go where the *--snip--* is shown in the setup code:

Verilog
```verilog
module Math_Examples();

  reg unsigned [3:0] i1_u4, i2_u4, o_u4;
  reg signed   [3:0] i1_s4, i2_s4, o_s4;

  reg unsigned [4:0] o_u5, i2_u5;
  reg signed   [4:0] o_s5, i1_s5, i2_s5;
  reg unsigned [5:0] o_u6;

  reg unsigned [7:0] o_u8, i_u8;
  reg signed   [7:0] o_s8;

  initial begin

    --snip--

    $finish();
  end

endmodule
```

VHDL
```vhdl
library ieee;
use ieee.std_logic_1164.all;
use ieee.numeric_std.all;
use std.env.finish;

entity Math_Examples is
end entity Math_Examples;

architecture test of Math_Examples is

  -- Takes input unsigned, returns string for printing
❶ function str(val : in unsigned) return string is
  begin
    return to_string(to_integer(val));
  end function str;

  -- Takes input signed, returns string for printing
❷ function str(val : in signed) return string is
  begin
    return to_string(to_integer(val));
  end function str;
```

```
    -- Takes input real, returns string for printing
❸ function str(val : in real) return string is
  begin
    return to_string(val, "%2.3f");
  end function str;

begin

  process is
    variable i1_u4, i2_u4, o_u4 : unsigned(3 downto 0);
    variable i1_u5, i2_u5, o_u5 : unsigned(4 downto 0);
    variable i1_s4, i2_s4, o_s4 : signed(3 downto 0);
    variable i1_s5, i2_s5, o_s5 : signed(4 downto 0);
    variable i1_u6, i2_u6, o_u6 : unsigned(5 downto 0);
    variable i1_u8, i2_u8, o_u8 : unsigned(7 downto 0);
    variable i1_s8, i2_s8, o_s8 : signed(7 downto 0);
    variable real1, real2, real3 : real;
  begin

    --snip--

    wait for 1 ns;
    finish;
  end process;

end test;
```

The skeleton for this testbench sets up a single initial (in Verilog) or process (in VHDL) block that runs once through. We'll fill in this block with examples later in the chapter. Notice that we've declared a number of signals using reg (in Verilog) and variable (in VHDL). This is the first time we've seen the variable keyword in VHDL: we need it so we can write blocking assignment statements in the testbench. See "Blocking vs. Non-Blocking Assignments" on page 214 for more information.

The examples in this chapter use a common naming scheme to quickly identify the data types and widths of the signals so you don't have to keep looking back at the signal definitions. The prefix i indicates an input to a math equation, while o indicates an output, the result of the math equation. In addition, we have the suffixes _u*N* and _s*N*, where u represents unsigned, s represents signed, and *N* represents the bit width of the signal. For example, o_s4 is a 4-bit-wide signed output. Establishing a scheme like this that makes it easier to remember data types and widths can be very helpful in your code, especially if there are many values in a single file.

Notice in the VHDL that we declare a custom function, str(), to help convert the outputs of our equations to strings for printing. This will save us a lot of typing in the examples later on. We actually define the function in three different ways, depending on the data type involved—because VHDL is strongly typed, we need to define all the supported function inputs so that the compiler knows which one to use. The first definition ❶ converts an unsigned value, the second ❷ converts a signed value, and the third ❸ converts a real value. This is an example of function *overloading*, a programming technique where a single function can have multiple implementations. Overloading is a somewhat advanced VHDL concept, but it's very useful. You can even overload normal VHDL operators like + and – with any implementation that you need, though I don't recommend doing so.

Now that we have our testbench set up, we're ready to start exploring math operations.

BLOCKING VS. NON-BLOCKING ASSIGNMENTS

Up to this point, we have been assigning all signals within our always, initial, and process blocks with <=, which is a non-blocking assignment. As we discussed in Chapter 4, this means that these statements execute at the same instant in time. Non-blocking assignments are what allow us to write several statements one after the other that will all execute on the same clock edge. These assignments are key to FPGA design; they are how we can write operations that occur in parallel, instead of serially. Remember, flip-flops that share the same clock are all updated on the same clock edge, all at once.

There's also such a thing as a *blocking assignment*, however, written with = in Verilog or := for VHDL variables, where the next line of code won't execute until the current line is finished running. Blocking assignments are probably familiar if you have experience with conventional programming languages. We're used to the idea that when we write two lines of code in a language like C or Python, the second line won't execute until the first line is done. In FPGA design, however, non-blocking assignments are the norm. They're the preferred way to generate sequential logic (flip-flops). When in doubt, stick to non-blocking assignments in your always, initial, and process blocks.

An exception is if you're writing a testbench, where blocking assignments are useful when data needs to be updated at an exact point during the simulation, especially for printing. To see why, consider this Verilog testbench code:

```
int test = 0;
initial begin
  test <= 7;
  $display("value is %d\n", test);
```

We change the value of test from 0 to 7, then use $display to print it out. But what value of test will be printed to the console? It's actually going to be 0, not 7. The line test <= 7; uses a non-blocking assignment, so it happens at the same time as the printing. Therefore, the value of test won't have updated yet when $display reads that value for printing. We can fix this issue by using a blocking assignment instead:

```
int test = 0;
initial begin
  test = 7;
  $display("value is %d\n", test);
```

Thanks to the blocking assignment (test = 7;), the simulation will wait to execute the $display line until the value of test has been updated. The value will now print out as 7, which is probably what was intended.

Another way you can fix this issue is to add a small delay between the non-blocking assignment and the next line, like this:

```
test <= 7;
#1;
$display("value is %d\n", test);
```

The short delay (#1) provides enough simulation time for the non-blocking assignment to complete before printing, so the value will be displayed as 7.

In the testbench examples throughout this chapter, we'll be updating and then printing out a lot of values. Rather than add a bunch of small delays throughout the testbench, we'll use blocking assignments to ensure the values are updated before printing. Keep in mind that (as mentioned previously) to use the blocking assignment in VHDL, we need to declare values using the variable keyword rather than the signal keyword.

Addition

Adding binary data works the same way you were taught to add numbers in elementary school: you add them one digit at a time, working from right to left. For example, here's how to add the binary numbers 0010 and 1000:

```
  0010
+ 1000
------
 01010
```

To arrive at the result, you simply go digit by digit, starting with the least significant bit, adding the digits in that column together. If you get $1 + 1 = 10$ in a column, then you write 0 at the bottom of the column and carry the 1 to the next digit to the left.

Notice that the result of adding two 4-bit numbers together is 5 bits wide. This is our first rule of FPGA math:

Rule #1   When adding, the result should be at least 1 bit bigger than the biggest input.

The extra bit is needed in the case where adding the most significant bit requires a carry operation. Without the extra bit, we'd be truncating the result, which could produce an incorrect answer. In our first example, dropping the most significant bit wouldn't make a difference, but consider this example where having that extra bit is critical:

```
   1001
+  1011
------
 10100
```

Here, the most significant bit of the result is a 1. Had we just assumed that the output width would be the same as the input widths, then we would have dropped this bit and gotten the wrong answer. Our result would have been 0100 instead of 10100.

Perhaps you've noticed that I haven't explicitly said what these binary numbers represent yet, and whether they're positive or negative. For example, is 1001 unsigned and equal to 9, or is it signed and equal to the two's complement of 9, which is –7 (invert the bits to get 0110, then add 1 to get 0111)? The reason I haven't specified this is because the representation of the binary data ultimately doesn't affect how the math is performed, as long as the inputs and outputs are sized appropriately. Whether 1001 represents +9 or –7, the addition operation will be performed the same way. Of course, we care if the result is positive or negative, but the implementation of the addition doesn't change depending on whether the data types are signed or unsigned. Let's revisit our first example and consider what happens when we assign it various signed and unsigned combinations. Here's the example again:

```
   0010
+  1000
------
 01010
```

If both addition inputs are declared as unsigned types, then we have $2 + 8 = 10$. Pretty simple. If both addition inputs are declared as signed, then the first input is still 2, but the second input is –8. (Invert the bits to get 0111, add 1 to get 1000, and apply the negative sign to get –8.) So now we have $2 + -8$, which should equal –6, but the result, 01010, is still 10. Something isn't right here!

The problem is that we're not performing sign extension on the inputs. *Sign extension* is the operation of increasing the number of bits of a binary number while preserving the number's sign and value. This operation is required when the inputs are signed. Without it, we'll get an incorrect answer, as you've just seen. To perform sign extension on a signed value, simply replicate the most significant bit. For example, 1000 becomes 11000, and 0010 becomes 00010. Let's try that math again, this time first applying sign extension to our inputs:

```
  00010
+ 11000
-------
  11010
```

Our inputs are still 2 and –8. (For the latter, invert the bits of 11000 to get 00111, add 1 to get 01000, and apply the negative sign to get –8.) The answer, 11010, is the signed equivalent of –6, which is exactly what we want. Sign extension was the critical step to ensure we got the expected answer.

Sign extension is useful for unsigned values, too. In fact, since VHDL is strongly typed, the inputs and outputs to an addition operation must all have exactly the same width. You can't, for example, add two 4-bit inputs to produce a 5-bit output; everything must be 5 bits. That means we should revisit Rule #1 and add a small modification:

> **Rule #1 (modification #1)** When adding, the result should be at least 1 bit bigger than the biggest input, *before sign extension*. Once sign extension is applied, the input and output widths should match exactly.

For unsigned values, sign extension simply means adding a 0 as the new most significant bit. For example, unsigned 1000 becomes 01000. The good news for those of you using Verilog is that the code performs sign extension automatically when you're adding numbers. If you're using VHDL, however, you'll need to sign-extend your inputs manually using the resize() function, as you'll see in the coming examples. Both approaches have their pros and cons. Verilog is easier if you know what you're doing, as there's less to worry about, but it also leaves more room for mistakes (for example, trying to store data in too small a signal). VHDL's extra steps can be more confusing for beginners, and it generates cryptic errors when the rules aren't followed. On the other hand, VHDL ensures that you've matched widths and types every step along the way, so there's less room for error in the end.

Let's summarize what we've learned with a few code examples. Add this code to your testbench where you saw the *--snip--* earlier:

Verilog

```
// Unsigned + Unsigned = Unsigned (Rule #1 violation)
i1_u4 = 4'b1001; // dec 9
i2_u4 = 4'b1011; // dec 11
o_u4  = i1_u4 + i2_u4;
$display("Ex01: %2d + %2d = %3d", i1_u4, i2_u4, o_u4);

// Signed + Signed = Signed (Rule #1 violation)
i1_s4 = 4'b1001; // dec -7
i2_s4 = 4'b1011; // dec -5
o_s4  = i1_s4 + i2_s4;
$display("Ex02: %2d + %2d = %3d", i1_s4, i2_s4, o_s4);

// Unsigned + Unsigned = Unsigned (Rule #1 fix)
i1_u4 = 4'b1001; // dec 9
i2_u4 = 4'b1011; // dec 11
o_u5  = i1_u4 + i2_u4;
$display("Ex03: %2d + %2d = %3d", i1_u4, i2_u4, o_u5);

// Signed + Signed = Signed (Rule #1 fix)
i1_s4 = 4'b1001; // dec -7
i2_s4 = 4'b1011; // dec -5
o_s5  = i1_s4 + i2_s4;
$display("Ex04: %2d + %2d = %3d", i1_s4, i2_s4, o_s5);
```

```
-- Unsigned + Unsigned = Unsigned (Rule #1 violation)
i1_u4 := "1001"; -- dec 9
i2_u4 := "1011"; -- dec 11
o_u4  := i1_u4 + i2_u4;
report "Ex01: " & str(i1_u4) & " + " & str(i2_u4) & " = " & str(c_u4);

-- Signed + Signed = Signed (Rule #1 violation)
i1_s4 := "1001"; -- dec -7
i2_s4 := "1011"; -- dec -5
o_s4  := i1_s4 + i2_s4;
report "Ex02: " & str(i1_s4) & " + " & str(i2_s4) & " = " & str(o_s4);

-- Unsigned + Unsigned = Unsigned (Rule #1 fix)
i1_u4 := "1001"; -- dec 9
i2_u4 := "1011"; -- dec 11
❶ i1_u5 := resize(i1_u4, i1_u5'length);
i2_u5 := resize(i2_u4, i2_u5'length);
o_u5  := i1_u5 + i2_u5;
report "Ex03: " & str(i1_u5) & " + " & str(i2_u5) & " = " & str(o_u5);

-- Signed + Signed = Signed (Rule #1 Fix)
i1_s4 := "1001"; -- dec -7
i2_s4 := "1011"; -- dec -5
i1_s5 := resize(i1_s4, i1_s5'length);
i2_s5 := resize(i2_s4, i2_s5'length);
o_s5  := i1_s5 + i2_s5;
report "Ex04: " & str(i1_s5) & " + " & str(i2_s5) & " = " & str(o_s5);
```

Here's the output:

```
# Ex01:  9 + 11 =    4
# Ex02: -7 + -5 =    4
# Ex03:  9 + 11 =   20
# Ex04: -7 + -5 =  -12
```

```
# ** Note: Ex01: 9 + 11 = 4
#    Time: 0 ns  Iteration: 0   Instance: /math_examples
# ** Note: Ex02: -7 + -5 = 4
#    Time: 0 ns  Iteration: 0   Instance: /math_examples
# ** Note: Ex03: 9 + 11 = 20
#    Time: 0 ns  Iteration: 0   Instance: /math_examples
# ** Note: Ex04: -7 + -5 = -12
#    Time: 0 ns  Iteration: 0   Instance: /math_examples
```

First, we have two situations (Ex01 and Ex02) where Rule #1 isn't followed. We're using 4-bit inputs and storing the result in a 4-bit output, and we're not performing any sign extension. In both of these examples, we get the wrong answer. In Ex01, we add two unsigned numbers, 9 and 11, but get 4 as a result. The problem here is that we're dropping the most significant bit, which would be worth 16. (Indeed, $4 + 16 = 20$, which is the answer we should be getting.) In Ex02, we add two signed numbers representing negative values, and again we get the wrong answer.

The fix is to store the result in a 5-bit output, which we do in both Ex03 and Ex04. We've satisfied Rule #1, so the math works correctly. Notice that in the Verilog version, the sign extension happens automatically: we can simply assign 4-bit inputs to a 5-bit output, for example by writing o_u5 = i1_u4 + i2_u4;. In VHDL, however, we must explicitly match input and output widths, while preserving the sign and value of each input. To do this, we call the resize() function ❶. We use the VHDL tick attribute 'length to reference the length of the output signal, as we did when we were performing type conversions. Again, this is more flexible than hardcoding the desired width by writing something like resize(i1_u4, 5).

Another tip for performing successful addition operations is to never mix signed and unsigned values. Your inputs and outputs should be of the same type; otherwise you might get an incorrect answer. This brings us to our second rule of FPGA math:

Rule #2   Match types among inputs and outputs.

With VHDL, it's easy to follow Rule #2 because it will throw an error if you try to do a math operation where one input is signed and the other is unsigned. For example, say you write this in your testbench to try to add a 4-bit unsigned value (i1_u4) to a 4-bit signed value (i2_s4):

```
o_u4 := i1_u4 + i2_s4;
```

You'll end up with an error message indicating the tool doesn't know how to interpret the + operator given those inputs:

```
** Error: testbench.vhd(49): (vcom-1581) No feasible entries for infix
operator '+'.
```

Verilog is much more lenient. It will happily let you perform that math operation, and it won't tell you that it's actually treating your signed input as unsigned. This can very possibly result in the wrong answer, so be careful to always match your data types in Verilog.

Subtraction

Subtraction isn't that different from addition. After all, subtraction is just an addition operation where one of the inputs is negative. In this sense, we've been doing subtraction all along; 2 + −8 is the same as 2 − 8. Likewise, you can think of something like 5 − 3 as 5 + −3 and approach it like an addition operation.

There's one thing to be careful with when subtracting two numbers, though: while you *could* use subtraction with unsigned inputs and outputs, I wouldn't recommend it. What happens if the result should be negative? For example, $3 - 5 = -2$, but if you try to store -2 into an unsigned data type, you won't get the correct result. This brings us to our next rule:

Rule #3   When subtracting, use signed inputs and outputs.

Even if you don't think the result of a subtraction will produce a negative number, you should use signed data types to be safe.

Because subtraction is really just negative addition, subtraction carries the same risk that you could truncate the result if the output isn't sized appropriately. Again, it's better to size up the output by 1 bit and to sign-extend your inputs before performing the math operation. This gives us a further modified Rule #1:

Rule #1 (modification #2)   When adding *or subtracting*, the result should be at least 1 bit bigger than the biggest input, before sign extension. Once sign extension is applied, the input and output widths should match exactly.

With those two rules in place, let's extend our Math_Examples testbench to take a look at some subtraction operations in Verilog and VHDL:

Verilog

```
// Unsigned - Unsigned = Unsigned (bad)
i1_u4 = 4'b1001; // dec 9
i2_u4 = 4'b1011; // dec 11
o_u5  = i1_u4 - i2_u4;
$display("Ex05: %2d - %2d = %3d", i1_u4, i2_u4, o_u5);

// Signed - Signed = Signed (fix)
i1_u4 = 4'b1001; // dec 9
i2_u4 = 4'b1011; // dec 11
❶ i1_s5 = i1_u4;
i2_s5 = i2_u4;
o_s5  = i1_s5 - i2_s5;
$display("Ex06: %2d - %2d = %3d", i1_s5, i2_s5, o_s5);
```

VHDL

```
-- Unsigned - Unsigned = Unsigned (bad)
i1_u4 := "1001"; -- dec 9
i2_u4 := "1011"; -- dec 11
i1_u5 := resize(i1_u4, i1_u5'length);
i2_u5 := resize(i2_u4, i2_u5'length);
o_u5  := i1_u5 - i2_u5;
report "Ex05: " & str(i1_u5) & " - " & str(i2_u5) & " = " & str(o_u5);

-- Signed - Signed = Signed (fix)
i1_u4 := "1001"; -- dec 9
i2_u4 := "1011"; -- dec 11
❷ i1_s5 := signed(resize(i1_u4, i1_s5'length));
i2_s5 := signed(resize(i2_u4, i2_s5'length));
o_s5  := i1_s5 - i2_s5;
report "Ex06: " & str(i1_s5) & " - " & str(i2_s5) & " = " & str(o_s5);
```

Here's the output:

```
# Ex05: 9 - 11 = 30
# Ex06: 9 - 11 = -2
```

VHDL

```
# ** Note: Ex05: 9 - 11 = 30
#    Time: 0 ns  Iteration: 0   Instance: /math_examples
# ** Note: Ex06: 9 - 11 = -2
#    Time: 0 ns  Iteration: 0   Instance: /math_examples
```

In Ex05, we're trying to calculate $9 - 11$ but we get a result of 30, clearly the wrong answer. The problem here is that we're using unsigned types for subtraction, which is a violation of Rule #3. We fix this in Ex06 by converting the input values from unsigned to signed data types. We also perform sign extension in the process, going from 4-bit inputs to 5-bit inputs. In the Verilog code, we handle the conversion by simply assigning the 4-bit unsigned signals to 5-bit signed signals ❶. Verilog takes care of the details automatically. VHDL makes us jump through a few more hoops. We first resize the input, which will sign-extend it, but the result of the resize operation is still an unsigned type, so we then explicitly cast it to a signed data type using signed() ❷. This is safe to do because we've already resized the signal, so the most significant bit will be 0. Therefore, the value after converting to a signed type won't be changed.

Multiplication

Multiplication also works similarly to addition; after all, a multiplication operation is just a series of repeated addition operations ($4 \times 3 = 4 + 4 + 4$). The first thing to consider when multiplying two inputs together is how to properly size the output bit width. This brings us to our next rule:

> **Rule #4** When multiplying, the output bit width must be at least the sum of the input bit widths (before sign extension).

This rule holds true for both signed and unsigned numbers. For example, say we're trying to multiply the unsigned inputs 111 and 11 (equivalent to 7×3). According to Rule #4, the output should be $3 + 2 = 5$ bits wide. You can try out the multiplication yourself to confirm this, using the same technique you learned in school for multiplying multidigit numbers—multiply each digit individually and add the results together:

```
    111
  ×  11
  ------
    111
+ 1110
  ------
  10101
```

The output, 10101 (equivalent to 21), is indeed 5 bits wide, which is what we expected. But what happens to this same multiplication if we treat our inputs and outputs as signed, rather than unsigned? In this case, we would have the equivalent of −1 × −1 in decimal, which should produce a result of +1, but signed 10101 in binary is equal to −11 in decimal. What's wrong here?

The problem is that we didn't sign-extend our inputs to match the width of our output (5 bits) before multiplying. If we do that, our inputs both become 11111, and the multiplication looks like this:

```
        11111
  ×     11111
  -----------
        11111
       111110
      1111100
     11111000
 + 111110000
  -----------
    000000001
```

Now we're getting 00000001, or really 00001 once we truncate the result to be 5 bits wide, which is +1 in decimal. Sign extension gives us the result we expect. Unlike with addition and subtraction, however, you don't actually need to perform this sign extension manually when multiplying numbers using Verilog or VHDL. The tools will handle this automatically; you simply need to size the output signal correctly.

VHDL helps with this too: it won't even let you compile the code if you disobey Rule #4 and fail to size the output of a multiplication correctly. With Verilog, you'll need to be more careful. It won't warn you if the output is the wrong size, and you could get an unexpected result. Let's add some examples of this to our Math_Examples testbench:

Verilog

```
// Unsigned * Unsigned = Unsigned (Rule #4 violation)
i1_u4 = 4'b1001; // dec 9
i2_u4 = 4'b1011; // dec 11
o_u4  = i1_u4 * i2_u4;
$display("Ex07: %2d * %2d = %3d", i1_u4, i2_u4, o_u4);

// Signed * Signed = Signed (Rule #4 violation)
i1_s4 = 4'b1000; // dec -8
i2_s4 = 4'b0111; // dec 7
o_s4  = i1_s4 * i2_s4;
$display("Ex08: %2d * %2d = %3d", i1_s4, i2_s4, o_s4);
```

VHDL

```
-- Unsigned * Unsigned = Unsigned
i1_u4 := "1001"; -- dec 9
i2_u4 := "1011"; -- dec 11
o_u4  := i1_u4 * i2_u4;
report "Ex07: " & str(i1_u4) & " * " & str(i2_u4) & " = " & str(o_u4);

-- Signed * Signed = Signed
i1_s4 := "1000"; -- dec -8
i2_s4 := "0111"; -- dec 7
```

```
o_s4  := i1_s4 * i2_s4;
report "Ex08: " & str(i1_s4) & " * " & str(i2_s4) & " = " & str(o_s4);
```

Here's the output:

```
# Ex07:  9 * 11 =  3
# Ex08: -8 *  7 = -8
```

VHDL

```
** Error (suppressible): testbench.vhd(89): (vcom-1272) Length of expected
is 4; length of actual is 8.
```

Verilog allows us to perform the math operation despite the fact that we're disobeying Rule #4 by multiplying 4 bits by 4 bits and storing the result in a 4-bit output. This produces incorrect results for both unsigned (Ex07) and signed (Ex08) input values. VHDL, on the other hand, won't even build this code; we get a nice descriptive error telling us that the tool is trying to assign an 8-bit-wide result to a 4-bit-wide variable, which isn't permitted. Let's add a few more examples to our testbench that fix these issues:

Verilog

```
// Unsigned * Unsigned = Unsigned (Rule #4 fix)
i1_u4 = 4'b1001; // dec 9
i2_u4 = 4'b1011; // dec 11
o_u8  = i1_u4 * i2_u4;
$display("Ex09: %2d * %2d = %3d", i1_u4, i2_u4, o_u8);

// Signed * Signed = Signed (Rule #4 fix)
i1_s4 = 4'b1000; // dec -8
i2_s4 = 4'b0111; // dec 7
o_s8  = i1_s4 * i2_s4;
$display("Ex10: %2d * %2d = %3d", i1_s4, i2_s4, o_s8);
```

VHDL

```
-- Unsigned * Unsigned = Unsigned
i1_u4 := "1001"; -- dec 9
i2_u4 := "1011"; -- dec 11
o_u8  := i1_u4 * i2_u4;
report "Ex09: " & str(i1_u4) & " * " & str(i2_u4) & " = " & str(o_u8);

-- Signed * Signed = Signed
i1_s4 := "1000"; -- dec -8
i2_s4 := "0111"; -- dec 7
o_s8  := i1_s4 * i2_s4;
report "Ex10: " & str(i1_s4) & " * " & str(i2_s4) & " = " & str(o_s8);
```

Here's the output:

Verilog

```
# Ex09:  9 * 11 =  99
# Ex10: -8 *  7 = -56
```

VHDL
```
# ** Note: Ex09: 9 * 11 = 99
#    Time: 0 ns  Iteration: 0   Instance: /math_examples
# ** Note: Ex10: -8 * 7 = -56
#    Time: 0 ns  Iteration: 0   Instance: /math_examples
```

In Ex09, we correct the problem in Ex07 by storing the output of multiplying two unsigned 4-bit values into an 8-bit signal. Similarly, Ex10 corrects the issue from Ex08 with signed values. Notice that we never have to sign-extend the inputs, in either Verilog or VHDL. The tools handle this automatically.

Multiplication by Powers of 2

There's a trick that we can use when multiplying numbers by a power of 2 (for example, 2, 4, 8, 16, 32, . . .). Rather than instantiating a bunch of multiplication logic, we can simply instantiate a shift register and perform a shift left operation. Shifting left by N bits is equivalent to multiplying by 2^N. For example, 0011 (3 in binary) shifted left 2 bits gives us 1100 (12 in binary). It's the same as calculating 3×4, or 3×2^2. This trick works for both signed and unsigned numbers. Let's try it out in our testbench:

Verilog
```
i_u8 = 3;
o_u8 = i_u8 << 1;
$display("Ex11: %d * 2 = %d",  i_u8, o_u8);
o_u8 = i_u8 << 2;
$display("Ex12: %d * 4 = %d",  i_u8, o_u8);
o_u8 = i_u8 << 4;
$display("Ex13: %d * 16 = %d", i_u8, o_u8);
```

VHDL
```
i1_u8 := to_unsigned(3, i1_u8'length);
o_u8 := shift_left(i1_u8, 1);
report "Ex11: " & str(i1_u8) & " * 2 = "  & str(o_u8);
o_u8 := shift_left(i1_u8, 2);
report "Ex12: " & str(i1_u8) & " * 4 = "  & str(o_u8);
o_u8 := shift_left(i1_u8, 4);
report "Ex13: " & str(i1_u8) & " * 16 = " & str(o_u8);
```

Here's the output:

Verilog
```
# Ex11:  3 * 2 =  6
# Ex12:  3 * 4 = 12
# Ex13:  3 * 16 = 48
```

VHDL
```
# ** Note: Ex11: 3 * 2 = 6
#    Time: 0 ns  Iteration: 0   Instance: /math_examples
# ** Note: Ex12: 3 * 4 = 12
#    Time: 0 ns  Iteration: 0   Instance: /math_examples
# ** Note: Ex13: 3 * 16 = 48
#    Time: 0 ns  Iteration: 0   Instance: /math_examples
```

We start with the decimal value 3 and shift left by 1, 2, and 4 bits to multiply it by 2, 4, and 16, respectively. In Verilog we perform the shift using the ≪ operator, and in VHDL we use the function shift_left(). Both take as an argument the number of bit positions to shift.

Shifting left is a simple and quick trick to save FPGA resources, but you don't necessarily need to write it out explicitly. It's likely that if you hardcode a multiplication by a power of 2, the synthesis tools will be smart enough to figure out that a left shift would take fewer resources.

Division

Unfortunately, division isn't nearly as simple an operation as addition, subtraction, or multiplication. Division comes with all sorts of messy complications, like remainders and fractions. In general, it's a good idea to avoid division inside your FPGA if you can. It's a resource-intensive operation, especially if you need that operation to run at high clock rates.

I once worked on a project that needed to add a division operation to an FPGA in the field. The FPGA was a very old part, and it simply couldn't fit it within the available resources. To accommodate the division operation we ended up having to upgrade to a higher-resource FPGA of the same family, which increased the cost of the hardware by over $1 million. I always think of that one extra operation as the million-dollar divide!

If you *must* divide numbers, there are a few ways to make the operation less resource-intensive. These include restricting yourself to dividing by powers of 2, using a precalculated table of answers, or breaking up the operation across multiple clock cycles.

Using Powers of 2

My best suggestion for reducing the overhead of dividing numbers inside an FPGA is to make the divisor a power of 2. Similar to how multiplication by a power of 2 can be efficiently performed with a shift left operation, division by a power of 2 can be performed efficiently with a shift right operation. Shifting right by N bits is equivalent to dividing by 2^N. Let's look at a few examples of this:

Verilog
```
i_u8 = 128;
o_u8 = i_u8 >> 1;
$display("Ex14: %d / 2 = %d",  i_u8, o_u8);
o_u8 = i_u8 >> 2;
$display("Ex15: %d / 4 = %d",  i_u8, o_u8);
o_u8 = i_u8 >> 4;
$display("Ex16: %d / 16 = %d", i_u8, o_u8);
```

VHDL
```
i1_u8 := to_unsigned(128, i1_u8'length);
o_u8 := shift_right(i1_u8, 1);
report "Ex14: " & str(i1_u8) & " / 2 = "  & str(o_u8);
o_u8 := shift_right(i1_u8, 2);
report "Ex15: " & str(i1_u8) & " / 4 = "  & str(o_u8);
o_u8 := shift_right(i1_u8, 4);
report "Ex16: " & str(i1_u8) & " / 16 = " & str(o_u8);
```

Here's the output:

Verilog
```
# Ex14: 128 / 2 = 64
# Ex15: 128 / 4 = 32
# Ex16: 128 / 16 =  8
```

VHDL
```
# ** Note: Ex14: 128 / 2 = 64
#    Time: 0 ns  Iteration: 0  Instance: /math_examples
# ** Note: Ex15: 128 / 4 = 32
#    Time: 0 ns  Iteration: 0  Instance: /math_examples
# ** Note: Ex16: 128 / 16 = 8
#    Time: 0 ns  Iteration: 0  Instance: /math_examples
```

Ex14 performs a shift right by 1, which in Verilog uses the >> operator and in VHDL uses the shift_right() function. This accomplishes a single divide by 2. To divide by 4, shift right by 2 bit positions, as in Ex15. Likewise, a right shift by 4 divides by 16, as in Ex16.

What happens when we don't have a number that's cleanly divisible by the power of 2 serving as the divisor? In this case, shifting right effectively accomplishes a division that's rounded down to the nearest integer. The next few examples illustrate how this works:

Verilog
```
i_u8 = 15;
o_u8 = i_u8 >> 1;
$display("Ex17: %d / 2 = %d", i_u8, o_u8);
o_u8 = i_u8 >> 2;
$display("Ex18: %d / 4 = %d", i_u8, o_u8);
o_u8 = i_u8 >> 3;
$display("Ex19: %d / 8 = %d", i_u8, o_u8);
```

VHDL
```
i1_u8 := to_unsigned(15, i1_u8'length);
o_u8 := shift_right(i1_u8, 1);
report "Ex17: " & str(i1_u8) & " / 2 = " & str(o_u8);
o_u8 := shift_right(i1_u8, 2);
report "Ex18: " & str(i1_u8) & " / 4 = " & str(o_u8);
o_u8 := shift_right(i1_u8, 3);
report "Ex19: " & str(i1_u8) & " / 8 = " & str(o_u8);
```

Here's the output:

Verilog
```
# Ex17: 15 / 2 =  7
# Ex18: 15 / 4 =  3
# Ex19: 15 / 8 =  1
```

VHDL
```
# ** Note: Ex17: 15 / 2 = 7
#    Time: 0 ns  Iteration: 0  Instance: /math_examples
# ** Note: Ex18: 15 / 4 = 3
#    Time: 0 ns  Iteration: 0  Instance: /math_examples
# ** Note: Ex19: 15 / 8 = 1
```

In Ex17, we try to perform 15 / 2. This should give us 7.5, but we have no way to represent the .5 part, so we end up rounding down to 7 instead. Thinking of this as a shift right, we went from 00001111 to 00000111. In Ex18, we try to take 15 / 4, which should be 3.75, but we drop the decimal places and just get 3. Finally, in Ex19 we get 15 / 8 = 1. This rounding might cause a problem if you're not expecting it, so be aware that this can happen when performing shift right operations.

Using a Precalculated Table

Another option for dividing two numbers is to precalculate the result for all possible input combinations. For example, if we're trying to divide any number 1 through 7 by any other number 1 through 7, we could create something like Table 10-3 inside the FPGA.

Table 10-3: Precalculated Table for Full Range of Division Inputs

	1	2	3	4	5	6	7
1	1.00	0.50	0.33	0.25	0.20	0.17	0.14
2	2.00	1.00	0.67	0.50	0.40	0.33	0.29
3	3.00	1.50	1.00	0.75	0.60	0.50	0.43
4	4.00	2.00	1.33	1.00	0.80	0.67	0.57
5	5.00	2.50	1.67	1.25	1.00	0.83	0.71
6	6.00	3.00	2.00	1.50	1.20	1.00	0.86
7	7.00	3.50	2.33	1.75	1.40	1.17	1.00

For this example, let's assume that each row represents a possible dividend, and each column represents a possible divisor. The value at the intersection of a given dividend and divisor is the corresponding quotient. As an example, to find the value in decimal for the fraction 5/6, go to row 5, then over to column 6 to get the value 0.83. To implement this in Verilog or VHDL, we could store this two-dimensional table in a 2D array. (You saw how 2D arrays work in the state machine project in Chapter 8.) The row input values provide one index, the column input values provide the second index, and the quotient is the value at those two indices. We're not actually performing any math here; we're just indexing into the correct result, which has been precalculated and stored in memory.

NOTE *If you're wondering how to represent decimal values like 0.50 and 0.33 inside an FPGA, good question! We'll explore this topic shortly.*

As the range of possible inputs grows, of course, we'll need a larger and larger table to store the possible outputs. Eventually, a single table could take up an entire block RAM, which are often 16Kb in size. Using a precalculated table in a block RAM guarantees that a single division calculation will take a single clock cycle, since we only need one clock cycle to read from the memory (as you learned when we discussed RAM back in Chapter 6). However, we can't read from multiple locations in the memory on the same clock cycle, so if we needed to do two divisions simultaneously, on the exact same clock cycle, we would need to instantiate a second copy of the precalculated table in another block RAM.

Block RAMs are usually valuable resources, so taking up a bunch of them for concurrent divisions doesn't scale very well. If the design will allow us to run the different divisions in consecutive clock cycles, rather than simultaneously, we could instead use a single table and time-share it. Time sharing a single resource would require arbitration of that resource, as we discussed in Chapter 7. We would have to create some arbiter that would only allow access to the block RAM table by one module at a time.

The solutions discussed up to this point assume we have just one clock cycle to get the result of a division operation. However, if we can wait multiple clock cycles for the result of a division operation, that allows us to use another option.

Using Multiple Clock Cycles

Another way to ease the burden of the synthesis tools when it comes to division is to create an algorithm that performs division in more than a single clock cycle, using simpler math operations such as addition and subtraction. At its heart, division is about calculating how many times one number fits into another number. You can accomplish this, for example, by adding the divisor to itself over and over until you've passed the value of the dividend, while counting the number of times you had to run that loop. Then you subtract the dividend to get the remainder.

There are various other techniques for performing division using simpler math operations. (Specific implementations are beyond the scope of this book; search the web for *division algorithms on FPGAs* if you want to learn more.) But of course, these methods only work if you're able to wait multiple clock cycles for the result. Using multiple clock cycles to produce a result is a bit different in this context than the pipelining example we discussed in Chapter 7, where we broke up a complex math operation across multiple clock cycles to meet timing. In that case, we were still able to get a result every clock cycle, but the outputs were delayed a few clock cycles from the inputs.

In this case, we don't know how many clock cycles the division algorithm will take to provide a result, so we can't rely on a result each clock cycle. Ultimately, it's a question of trading lower resource utilization for more clock cycles. If you really need to get the result of a division operation on every single clock cycle, you'll have to use one of the previously discussed division techniques.

How FPGAs Implement Math Operations

With all the operations we've discussed so far, we've only looked at how the math works, without really considering how the operations are implemented inside an FPGA. There are various FPGA components that may be involved, depending on the specific operation performed. If you take an introductory digital electronics course, you might learn about *half adders* and *full adders*, digital circuits that combine various logic gates (like XOR and AND) to perform addition operations. It's a fascinating subject, but in the end you might be frustrated to find that you don't need to know how these circuits work to be able to do math with modern FPGA code. You'll never need to instantiate a full adder component by manually typing out all the necessary logic operations if you're just adding two numbers together. Instead, you just use the + operator in Verilog or VHDL, like you would in any other programming language, and trust the synthesis tools to handle the implementation.

The tools will likely place addition and subtraction operations into basic LUTs. For multiplication, the tools will use flip-flops for the shift left approach, or LUTs or DSP blocks (if available) for more complicated calculations. As discussed in Chapter 9, DSP blocks are useful for accelerating large multiply–accumulate operations without utilizing a lot of LUT logic. Finally, division will require registers for the shift right approach, block RAMs for the precalculated table approach, or LUTs.

There's more to math than just addition, subtraction, multiplication, and division, however. Look at your calculator and consider all the operations we haven't discussed: sine, cosine, square root, and more. It's certainly possible to run these operations on an FPGA, but it gets complicated and is beyond the scope of this book. If you're interested in learning more, there are dedicated algorithms that you can instantiate for these, such as a Coordinate Rotation Digital Computer (CORDIC). Search GitHub for *FPGA CORDIC* and you'll find many examples.

In addition to actually implementing more complicated math operations on your FPGA, if you have the option it might be worth sending the inputs to a dedicated processor to perform the calculations, and then returning the result back to the FPGA logic. We'll discuss floating- versus fixed-point arithmetic in the next section, but processors are much more capable of performing floating-point arithmetic than FPGAs. This processor can be a dedicated component external to the FPGA, or it can be internal to the FPGA itself. If it's internal, it's referred to as either a hard-core processor or a soft-core processor, depending on if it's a dedicated piece of silicon or not.

Many modern FPGAs have internal hard ARM cores. This type of component with FPGA logic and a dedicated processor is often referred to as a *system on a chip (SoC)*. You can send the operations from the FPGA LUT/flip-flop logic into the ARM core for processing, and it will perform whatever operation is required and return the result. This solution is more about handling data than performing math, since you'll likely need to set up FIFOs for each of the inputs and outputs. Working with a separate processor is an advanced topic, but it can be very valuable in higher-end applications.

Working with Decimals

So far we've been working with integers, but there are many applications where you'll need your FPGA to operate on numbers with a decimal component. In this section, we'll examine how to do math using non-integers. To begin with, we need to consider how fractional numbers are actually represented using binary digits. There are two possible systems to choose from: floating point and fixed point.

The vast majority of mathematical operations within electronic devices use *floating-point* arithmetic, since most CPUs are designed to handle floating-point numbers. The key to floating point is that the *radix* (the decimal separator) "floats," depending on how much precision is needed. We won't go into detail about how exactly this works, but the bottom line is that with 32 bits you can represent an enormous range of values, with varying precision; you can represent very small numbers with high precision, or very large numbers with less precision. *Fixed-point* arithmetic, on the other hand, has a fixed radix, meaning there are a fixed number of integer places and a fixed number of decimal places.

FPGAs *can* perform floating-point operations, but they often require more resources than fixed-point operations. Most FPGA math is therefore done with fixed-point arithmetic, so that will be our focus for the rest of the chapter.

To illustrate how fixed-point representation works, let's take an example. Say we have 3 bits allotted for representing a number inside our FPGA. We've been assuming up to this point that each bit change will be worth one integer value. For example, going from 001 to 010 means that we go from 1 to 2. But we've just arbitrarily decided that each bit is worth one integer. We could just as easily decide that a single bit is worth something else, for example 0.5. In that case, 001 would be equivalent to 0.5, 010 would be 1.0, 011 would be 1.5, and so on. We now have a fixed-point system where the rightmost bit represents the decimal component of the number and the other two bits represent the integer component. We can also interpret the bits in other ways to give us different fixed-point representations. Table 10-4 shows the most common decimal interpretations of 3 unsigned bits.

Table 10-4: 3-Bit Unsigned Fixed-Point Possibilities

Bits	U3.0	U2.1	U1.2	U0.3
000	0	0	0	0
001	1	0.5	0.25	0.125
010	2	1.0	0.50	0.250
011	3	1.5	0.75	0.375
100	4	2.0	1.00	0.500
101	5	2.5	1.25	0.625
110	6	3.0	1.50	0.750
111	7	3.5	1.75	0.875

The headings in Table 10-4 use a modified version of *Q notation*, which is a way to specify the parameters of a binary fixed-point number format. In Q notation, for example, Q1.2 indicates that 1 bit is being used for the integer portion of a number and 2 bits are being used for the fractional portion. Standard Q notation assumes the values are signed; however, in FPGAs it's very common to have unsigned and signed values. That's why I prefer a notation that specifies if the values are signed (S) or unsigned (U) using the leading character. Thus, S3.1 indicates a signed value with 3 integer bits and 1 fractional bit, and U4.8 indicates an unsigned value with 4 integer bits and 8 fractional bits.

In Table 10-4, the U3.0 column is what we're used to; all 3 bits are allotted to the integer portion of the number, so we only have whole numbers. Let's consider the next column, U2.1. It's unsigned, with 2 bits for the integer component and 1 bit for the decimal component. This means the integer part can be in the range 00, 01, 10, 11, and the decimal part can be in the range 0 or 1. To figure out what possible values that represents, simply take the original U3.0 value and divide it by 2. For example, 111 is 7 in U3.0, but in U2.1 it's 3.5 (7 / 2 = 3.5). In general, when there are N bits allotted to the fractional portion of the number, you divide the integer representation by 2^N to determine the fixed-point value. Thus, 111 in U0.3 is $7 / 2^3 = 7 / 8 = 0.875$.

In Table 10-4 we treated all the values as unsigned. Table 10-5 shows the most common possibilities for interpreting the same 3 bits when we use signed data types.

Table 10-5: 3-Bit Signed Fixed-Point Possibilities

Bits	S3.0	S2.1	S1.2	S0.3
000	0	0	0	0
001	1	0.5	0.25	0.125
010	2	1.0	0.50	0.250
011	3	1.5	0.75	0.375
100	–4	–2.0	–1.00	–0.500
101	–3	–1.5	–0.75	–0.375
110	–2	–1.0	–0.50	–0.250
111	–1	–0.5	–0.25	–0.125

The S3.0 column shows the same signed whole number values we saw earlier in the chapter, in Table 10-1. We can generate the remaining columns by dividing the values in the S3.0 column by 2 for S2.1, by 4 for S1.2, and by 8 for S0.3.

Here's the critical thing about working with fixed-point numbers: when you're performing operations on binary data, the behavior of the binary operation doesn't change based on its representation. Addition, subtraction, multiplication, and division all work exactly the same way as before, when we were treating the numbers as integers. However, there are a few more rules that need to be established to get the correct answer with fixed-point values.

You'll notice for the remainder of this chapter that I make an effort to keep track of the decimals in the code examples. I find it very helpful to add comments recording the width of the math operations in my Verilog or VHDL code. For example, when adding two 3-bit numbers together to get a 4-bit result, I'll include a comment like `// U2.1 + U2.1 = U3.1` so I know the decimal and integer widths. This is particularly useful when there are several math operations chained one after another, where the widths along the way might be changing.

Adding and Subtracting with Fixed Point

When performing addition or subtraction with fixed-point decimals, the actual process doesn't change. The data is still just binary. There's another rule that we must apply when we have decimals involved, however:

Rule #5   When adding or subtracting, the decimal widths must match.

The number of places to the right of the decimal point determines the value, or *weight*, of each bit, so if you try to add or subtract two inputs with different decimal bit widths—for example, a U3.1 input and a U4.0 input—you'll get a wrong answer. We can see that in the following code:

Verilog
```
// U3.1 + U4.0 = U4.1 (Rule #5 violation)
i1_u4 = 4'b0011;
i2_u4 = 4'b0011;
o_u5  = i1_u4 + i2_u4;
❶ $display("Ex20: %2.3f + %2.3f = %2.3f", i1_u4/2.0, i2_u4, o_u5/2.0);
```

VHDL
```
-- U3.1 + U4.0 = U4.1 (Rule #5 violation)
i1_u4 := "0011";
i2_u4 := "0011";
i1_u5 := resize(i1_u4, i1_u5'length);
i2_u5 := resize(i2_u4, i2_u5'length);
o_u5  := i1_u5 + i2_u5;
❶ real1 := real(to_integer(i1_u5)) / 2.0;
real2 := real(to_integer(i2_u5));
real3 := real(to_integer(o_u5)) / 2.0;
report "Ex20: " & str(real1) & " + " & str(real2) & " = " & str(real3);
```

Here's the output:

Verilog
```
# Ex20: 1.500 + 3.000 = 3.000
```

VHDL
```
# ** Note: Ex20: 1.500 + 3.000 = 3.000
#    Time: 0 ns  Iteration: 0  Instance: /math_examples
```

Ex20 shows the effect of not obeying Rule #5. Here we're attempting to add a U3.1 to a U4.0. This is going to cause a problem because the weight of the bits being added together isn't matched. Indeed, the printout tells us that $1.5 + 3 = 3$, so something has clearly gone wrong.

Notice that we've divided the input U3.1 and the output U5.1 by 2.0 to print out these fixed-point values correctly ❶. For Verilog, we can simply do the division on the unsigned input and use %f to format the result like a float. In VHDL, the conversion is a bit more complicated. First we need to switch to the real data type, which is used for numbers with decimals, and then we can divide by 2.0 for printing.

To fix this example, we need to adjust one of the inputs so it has the same number of decimal bits as the other input. We can either change the first input from U3.1 to U4.0 to match the second input, or change the second input from U4.0 to U4.1. In the following code, we try both options:

Verilog

```
// Convert U3.1 to U4.0
// U4.0 + U4.0 = U5.0 (Rule #5 fix, using truncation)
i1_u4 = 4'b0011;
i2_u4 = 4'b0011;
❶ i1_u4 = i1_u4 >> 1; // Convert U3.1 to U4.0 by dropping decimal
o_u5  = i1_u4 + i2_u4;
$display("Ex21: %2.3f + %2.3f = %2.3f", i1_u4, i2_u4, o_u5);

// Or Convert U4.0 to U4.1
// U3.1 + U4.1 = U5.1 (Rule #5 fix, using expansion)
i1_u4 = 4'b0011;
i2_u4 = 4'b0011;
❷ i2_u5 = i2_u4 << 1;
o_u6  = i1_u4 + i2_u5;
$display("Ex22: %2.3f + %2.3f = %2.3f", i1_u4/2.0, i2_u5/2.0, o_u6/2.0);
```

VHDL

```
-- Convert U3.1 to U4.0
-- U4.0 + U4.0 = U5.0 (Rule #5 fix, using truncation)
i1_u4 := "0011";
i2_u4 := "0011";
❶ i1_u4 := shift_right(i1_u4, 1); -- Convert U3.1 to U4.0
i1_u5 := resize(i1_u4, i1_u5'length);
i2_u5 := resize(i2_u4, i2_u5'length);
o_u5  := i1_u5 + i2_u5;

real1 := real(to_integer(i1_u5));
real2 := real(to_integer(i2_u5));
real3 := real(to_integer(o_u5));
report "Ex21: " & str(real1) & " + " & str(real2) & " = " & str(real3);

-- Or Convert U4.0 to U4.1
-- U3.1 + U4.1 = U5.1 (Rule #4 fix, using expansion)
i1_u4 := "0011";
i2_u4 := "0011";
i1_u6 := resize(i1_u4, i1_u6'length); -- expand for adding
i2_u6 := resize(i2_u4, i2_u6'length); -- expand for adding
❷ i2_u6 := shift_left(i2_u6, 1); -- Convert 4.0 to 4.1
o_u6  := i1_u6 + i2_u6;
```

```
real1 := real(to_integer(i1_u6)) / 2.0;
real2 := real(to_integer(i2_u6)) / 2.0;
real3 := real(to_integer(o_u6)) / 2.0;
report "Ex22: " & str(real1) & " + " & str(real2) & " = " & str(real3);
```

Here's the output:

```
# Ex21: 1.000 + 3.000 = 4.000
# Ex22: 1.500 + 3.000 = 4.500
```

```
# ** Note: Ex21: 1.000 + 3.000 = 4.000
#    Time: 0 ns  Iteration: 0  Instance: /math_examples
# ** Note: Ex22: 1.500 + 3.000 = 4.500
#    Time: 0 ns  Iteration: 0  Instance: /math_examples
```

In Ex21, we convert the U3.1 to a U4.0, effectively dropping the decimal point. We do this using a 1-bit shift to the right ❶. But consider the effect of this: we're eliminating the least significant bit, and if that bit has a 1 in it, then we're dropping that data. Essentially, we're performing a rounding operation to the next lowest integer. We can see that our first input was originally 1.5, but after dropping the decimal point it's 1.0. The math is correct, $1.0 + 3.0 = 4.0$, but we've truncated our input.

Ex22 shows a better solution that retains the precision of all inputs. Rather than shifting the first input to the right, we shift the second input to the left ❷. This pads the least significant bit with a 0, converting our second input from U4.0 to U4.1. Notice that this means the second input now occupies a total of 5 bits. We need to be sure to resize it, or we could end up losing the data in the most significant bit during the shift left. Additionally, our output now has to be 6 bits so we don't violate Rule #1.

Now that the decimal widths of the two inputs are matched with no loss of precision, we're able to successfully calculate that $1.5 + 3.0 = 4.5$. Expanding your inputs to match is the best solution if you don't want to round any of the decimal values.

NOTE *Subtracting fixed-point numbers works with all the same rules as addition, so we won't consider an example here. Follow the rules introduced in this chapter, and your subtraction operations will work as expected.*

Multiplying with Fixed Point

Multiplication with fixed-point numbers doesn't require any shifting to match the decimal widths. Instead, we can simply multiply the two inputs together as they are, provided we keep track of the input widths and size the output appropriately. We already have a rule for multiplication:

Rule #4   When multiplying, the output bit width must be at least the sum of the input bit widths (before sign extension).

Now we need to add another rule to account for fixed-point numbers:

Rule #6   When multiplying fixed-point numbers, add the integer component bit widths and the decimal component bit widths of your inputs separately to get the output format.

For example, if you're trying to multiply a U3.5 by a U1.7, the result is formatted as a U4.12. We determine this by adding the integer components (3 + 1 = 4) and the decimal components (5 + 7 = 12), and putting them together to get the output width format. It works the same way for signed values, so S3.0 × S2.4 = S5.4. Notice that we're still obeying Rule #4 as well, since the output width will be the sum of the input widths. It's just that the integer and decimal components are treated separately.

Let's take a look at some examples in Verilog and VHDL:

Verilog
```
// U2.2 * U3.1 = U5.3
i1_u4 = 4'b0101;
i2_u4 = 4'b1011;
o_u8  = i1_u4 * i2_u4;
$display("Ex23: %2.3f * %2.3f = %2.3f", i1_u4/4.0, i2_u4/2.0, o_u8/8.0);

// S2.2 * S4.0 = S6.2
i1_s4 = 4'b0110;
i2_s4 = 4'b1010;
o_s8  = i1_s4 * i2_s4;
$display("Ex24: %2.3f * %2.3f = %2.3f", i1_s4/4.0, i2_s4, o_s8/4.0);
```

VHDL
```
-- U2.2 * U3.1 = U5.3
i1_u4 := "0101";
i2_u4 := "1011";
o_u8  := i1_u4 * i2_u4;

real1 := real(to_integer(i1_u4)) / 4.0;
real2 := real(to_integer(i2_u4)) / 2.0;
real3 := real(to_integer(o_u8))  / 8.0;
report "Ex23: " & str(real1) & " * " & str(real2) & " = " & str(real3);

-- S2.2 * S4.0 = S6.2
i1_s4 := "0110";
i2_s4 := "1010";
o_s8  := i1_s4 * i2_s4;

real1 := real(to_integer(i1_s4)) / 4.0;
real2 := real(to_integer(i2_s4));
real3 := real(to_integer(o_s8))  / 4.0;
report "Ex24: " & str(real1) & " * " & str(real2) & " = " & str(real3);
```

Here's the output:

Verilog
```
# Ex23: 1.250 * 5.500 = 6.875
# Ex24: 1.500 * -6.000 = -9.000
```

VHDL
```
# ** Note: Ex23: 1.250 * 5.500 = 6.875
#    Time: 0 ns  Iteration: 0  Instance: /math_examples
# ** Note: Ex24: 1.500 * -6.000 = -9.000
#    Time: 0 ns  Iteration: 0  Instance: /math_examples
```

In Ex23, we're multiplying a U2.2 by a U3.1 to get a result that's a U5.3. We can see in the printout that the answer is correct: $1.25 \times 5.5 = 6.875$. As with the addition examples, notice that we have to divide the values to print them out correctly. We divide the U2.2 by 4, the U3.1 by 2, and the U5.3 by 8. In Ex24, we use the same technique to multiply signed values. We're multiplying 1.5 by −6.0 to get −9.0, which is represented with $S2.2 \times S4.0 = S6.2$.

Summary

Since FPGAs are known for being able to perform many calculations at fast clock rates and in parallel, many common FPGA applications call for using addition, subtraction, multiplication, and division. Inside your FPGA, these operations may involve LUTs, shift registers, or DSP blocks. More important than knowing exactly how the operations are implemented, however, is understanding how the inputs and outputs are stored and what those binary digits represent when you're writing your Verilog or VHDL code. Are they signed or unsigned? Integers or fixed point?

Over the course of this chapter, we've developed a set of rules for successfully performing FPGA math operations and interpreting the results. They are:

Rule #1   When adding or subtracting, the result should be at least 1 bit bigger than the biggest input, before sign extension. Once sign extension is applied, the input and output widths should match exactly.

Rule #2   Match types among inputs and outputs.

Rule #3   When subtracting, use signed inputs and outputs.

Rule #4   When multiplying, the output bit width must be at least the sum of the input bit widths (before sign extension).

Rule #5   When adding or subtracting, the decimal widths must match.

Rule #6   When multiplying fixed-point numbers, add the integer component bit widths and the decimal component bit widths of your inputs separately to get the output format.

These rules don't capture every nuance of performing math in FPGAs, but they cover the major details that you need to get right. If you follow these six rules, it's much more likely that you'll get the correct answer from your calculations. Whenever you're working with math, adding tests will help to ensure things are working as you expect.

11

GETTING DATA IN AND OUT WITH I/O AND SERDES

Throughout this book, we've focused on the internals of FPGAs, and that's typical of the FPGA design process. FPGA design largely centers around writing Verilog or VHDL code targeting internal components like flip-flops, LUTs, block RAMs, and DSP blocks. But what's going on at the edge of the device, where data enters and exits the FPGA?

There's a surprising amount of complexity involved in getting data into and out of an FPGA. In my experience, this is where most of the trickier FPGA design problems occur. Understanding how the *input/output (I/O)* works will help you tackle those problems. You'll be able to spend less time worrying about external interfaces, and more time solving the internal task at hand.

Working with I/O is where the boundary between being a "software person" and a "hardware person" lies. You need to understand the details of the electrical signals that you're interfacing to in order to configure the FPGA pins correctly. What voltage do they operate at? Are the signals single-ended or differential? (And what does that even mean?) How can you use double data rate or a serializer/deserializer to send data at very high speeds? This chapter answers these questions and more. Even if you don't have an electrical engineering background, you'll learn the fundamentals of interfacing FPGAs to the outside world.

Working with GPIO Pins

Most pins on the FPGA are *general purpose input/output (GPIO)* pins, meaning they can function as a digital input or output. We've used these pins in the book's projects to take in signals from push buttons and output signals to light up LEDs, but we haven't worried about the details of how this actually works. In this section, we'll look at how GPIO pins interface with an FPGA and how they can be made to input data, output data, or both.

When I was first getting into FPGA design, I had no idea of the nuances involved in pin configuration. There are many knobs to turn and settings to play with. Having a thorough understanding of your GPIO pins is important, especially for high-speed designs, because maintaining signal integrity and performance throughout your design starts at the pins.

I/O Buffers

GPIO pins interface with an FPGA through *buffers*, electronic circuit elements that isolate their input from their output. These buffers are what allow you to configure some pins as inputs and others as outputs. As you'll see soon, they even allow you to toggle a pin between input and output while the FPGA is running. Figure 11-1 shows a simplified block diagram of a GPIO pin interface on an Intel FPGA, to illustrate how a buffer serves as an intermediary between the pin and the internal FPGA logic.

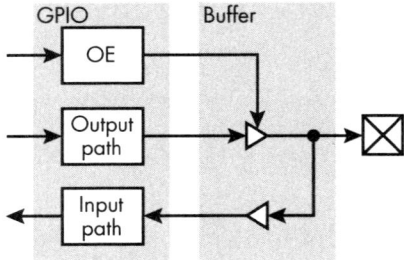

Figure 11-1: A simplified GPIO block diagram

The box on the right-hand side of the image (with the X inside it) represents the physical pin. Immediately to the left of the pin is a block labeled Buffer, which represents the I/O buffer. It contains two main components, represented by triangles. The triangle pointing to the right is the *output buffer*; it pushes data out to the pin. The triangle pointing to the left is the *input buffer*; it sends data from the pin into the FPGA.

On the far left of the diagram is a block labeled GPIO, representing the internal FPGA logic that interacts directly with the pin via the buffers. The main path to notice here is OE, which stands for *output enable*. This turns the output buffer on or off to control whether the pin will function as an output or an input. When OE is high, the output buffer will drive the pin with whatever data is present on the output path. If data on the output path is low, the pin will be low, and if data on the output path is high, the pin will be high. When OE is low, the pin is configured as an input, so the output buffer stops passing its input to its output. At this point the buffer's output becomes *high impedance* (also called *hi-Z* or *tri-state*), meaning it will accept very little current. A high-impedance output buffer no longer affects anything happening on the pin. Instead, the pin's state is governed by whatever external signal is coming in. The input buffer is free to read that signal and pass it along to the input path for use inside the FPGA.

Table 11-1 shows a truth table for an output buffer, summarizing this behavior.

Table 11-1: Truth Table for an Output Buffer

Input	OE	Output
0	0	Z
1	0	Z
0	1	0
1	1	1

Looking at this table, we can see that when OE is high, the value on the buffer's input is simply passed to its output. However, when OE is low, the buffer's output is high impedance (conventionally represented by a *Z*), regardless of the value on the input.

In the projects in this book, we've defined the input and output signals at the top level of the design code. Inputs are represented with the keyword input (Verilog) or in (VHDL), while outputs are indicated by the keyword output (Verilog) or out (VHDL). When building the FPGA, the synthesis tools see which signals are defined for each direction and set up the buffers accordingly. If the signal is an input, OE will be set low. If the signal in an output, OE will be set high. Then, during the place and route process, the physical constraints file maps the signals to the specific pins on the FPGA. This is how GPIO pins get configured as dedicated input or output pins for your design.

Bidirectional Data for Half-Duplex Communication

While most pins in a design are typically fixed as either input or output, a GPIO pin can be configured to be *bidirectional*, meaning it can switch between functioning as input and output within the same design. When the FPGA needs to output data through the bidirectional pin, it drives the OE signal high, then puts the data to transmit onto the output path. When the FPGA needs to receive data as input through the bidirectional pin, it drives OE low. This puts the output buffer into tri-state (high impedance), enabling the FPGA to listen to the data on the pin and pass it to the input path. When a pin is configured to be bidirectional like this, it's acting as a *transceiver*, as opposed to just a transmitter or just a receiver.

Bidirectional pins are useful for *half-duplex* communication, where two devices exchange data using a single, shared transmission line (one pin). Either device can serve as a transmitter, but only one device can transmit at a time, while the other receives. This is in contrast to *full-duplex* communication, where both devices can transmit and receive data at the same time. Full-duplex communication requires two transmission lines (two pins), one for sending data from device 1 to device 2 and the other for sending data from device 2 to device 1, as opposed to the single transmission line of half-duplex communication.

A common example of half-duplex communication is a two-way radio. The speaker is only able to transmit when they hold down the button on the radio. When the speaker is transmitting, the listener is unable to transmit, so the speaker and listener must agree whose turn it is to talk. This is why people always say "Over" in the movies when they're talking on walkie-talkies; it's a signal that the speaker is done talking and the listener is now free to respond.

With a physical wire, if the two sides don't take turns sharing the communication channel, then there can be a data collision. This collision can corrupt the data, so nobody receives anything. To avoid this the devices must agree on a *protocol*, a set of rules governing communication. The protocol determines how a device can initiate a transaction, establishes well-defined locations in time for other devices on the line to talk back (the equivalent of saying "Over"), and so on. Some protocols are even able to handle data collisions by detecting when data is corrupted and resending the corrupted data, though this requires additional complexity.

Half-duplex communication is usually more complicated than using dedicated transmit and receive channels, but it's still quite common. *I2C* (or I²C, pronounced "eye-squared-see" or "eye-two-see" and short for *inter-integrated circuit*), for example, is a widely used half-duplex protocol. Countless unique integrated circuits—including ADCs, DACs, accelerometers, gyroscopes, temperature sensors, microcontrollers, and many others—use I2C to communicate, since it's relatively simple to implement and, thanks to its half-duplex nature, requires a very low pin count. Only two pins, clock and data, are used in I2C, which is why you may also see it referred to as *TWI (two-wire interface)*.

A Bidirectional Pin Implementation

Let's look at how to code a bidirectional pin using Verilog or VHDL. As you examine this code, refer to Figure 11-2 to see how the signals in the code match the block diagram from Figure 11-1:

Verilog ❶ `module bidirectional(inout io_Data,`
`    --snip--`
❷ `assign w_RX_Data = io_Data;`
❸ `assign io_Data    = w_TX_En ? w_TX_Data : 1'bZ;`
`    --snip--`

VHDL `entity bidirectional is`
❶ `port (io_Data : inout std_logic,`
`    --snip--`
❷ `w_RX_Data <= io_Data;`
❸ `io_Data    <= w_TX_Data when w_TX_En = '1' else 'Z';`
`    --snip--`

We declare the bidirectional pin (io_Data) with the keyword inout in both Verilog and VHDL ❶. At this point we can imagine that we're at the pin, as indicated by the label io_Data in Figure 11-2. We'll need to map this signal to one of the FPGA's pins in our physical constraints file. For the input functionality, we simply use an assignment to drive w_RX_Data with the data from the pin ❷. On the output side, we selectively enable the output buffer using the signal w_TX_En ❸. We use the ternary operator in Verilog or a conditional assignment in VHDL. The data driven onto io_Data will either be w_TX_Data or high impedance (indicated by 1'bZ in Verilog or 'Z' in VHDL), depending on the state of the output enable signal (w_TX_En). This code pattern is very common for bidirectional data. Synthesis tools are smart enough to recognize it and infer an I/O buffer.

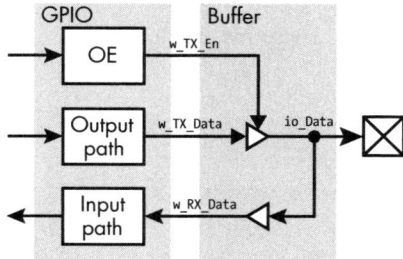

Figure 11-2: A labeled bidirectional interface

One thing you might notice is that any data driven out on w_TX_Data will be received on w_RX_Data, since they're connected together through io_Data. You'll need to address this elsewhere in the code by telling your receiver to ignore any data on io_Data when w_TX_En is high. Otherwise, your FPGA will be hearing itself talk.

Electrical Characteristics

There are many different electrical characteristics that you can specify for each individual GPIO pin. We're going to talk about three: operating voltage, drive strength, and slew rate. We'll also look at the electrical differences between single-ended and differential data transmission.

As you read, keep in mind that these aren't the only pin settings you can control. For example, you also may be able to set pins to be open drain, include a pull-up or pull-down resistor or a termination resistor, and much more. The I/O of your FPGA can be configured in many, many ways, depending on which GPIO properties are built into the device itself. If you need to implement anything other than simple signal interfaces, it's worth exploring the relevant datasheets to ensure you're working correctly with your I/O buffers. All of the specific information about your FPGA's I/O can usually be found in the I/O user guide, which is a great reference for details on what types of electronics your FPGA is capable of interfacing to.

Operating Voltage

The *operating voltage* specifies what voltage the pin will be driven to for a logic 1 output and sets the expected voltage for a logic 1 input. Most commonly, FPGA pins use 0 V to represent a 0 and 3.3 V to represent a 1. This standard is called *LVCMOS33* (LVCMOS is short for *low-voltage complementary metal oxide semiconductor*). Another standard you might come across is 0 V to represent a 0 and 5 V to represent a 1. This is called *TTL*, short for *transistor–transistor logic*. TTL is less common in FPGAs these days, since many don't allow voltages as high as 5 V internally. There's also the LVCMOS25 standard, which uses 0 V to represent a 0 and 2.5 V to represent a 1.

LVCMOS33, LVCMOS25, and TTL are all examples of *single-ended* I/O standards, meaning the signals involved are referenced to ground. As you'll see soon, there are also *differential* standards, where the signals aren't referenced to ground. There are many more single-ended standards than the three I've mentioned. A typical FPGA supports about a dozen single-ended I/O standards.

One important note about setting your operating voltage is that all signals on a single bank need to be at the same operating voltage. A *bank* is a group of pins that all operate with a common reference voltage, usually called *VCCIO*. You might have eight banks on your FPGA, and each bank can use a unique operating voltage. For example, you might configure bank 1 to use 1.8 V, bank 2 to use 3.3 V, and bank 3 to use 2.5 V. What's critical is that all the pins within a single bank are operating at the same voltage. You can't put an LVCMOS33 pin on the same bank as an LVCMOS25 pin, because the former requires a VCCIO of 3.3 V while the latter requires a VCCIO of 2.5 V. When doing your schematic review, always check to make sure that the signals on each bank share the same reference voltage. If you try to mix voltages in the same bank, the place and route tool will likely generate an error, or at least a very strong warning.

Drive Strength

The *drive strength* of a pin determines how much current (in milliamps, or mA) can be driven into or out of the pin. For example, a pin set to an 8 mA drive strength will be capable of sinking or sourcing up to 8 mA of current. The drive strength can be changed on an individual pin basis and should be high enough to match the needs of the circuit you're interfacing to. Most often, the drive strength settings can be left at the default for all of the pins on your FPGA. Unless you have some high current needs, it's unlikely you'll need to modify the default settings.

Slew Rate

The *slew rate* sets the rate of change allowed for an output signal. It's usually specified in qualitative terms, such as *fast, medium,* or *slow.* The slew rate setting affects how quickly a pin can change from a 0 to a 1 or from a 1 to a 0. Like drive strength, the slew rate can often be left at the default setting for each of your pins. The exception is if you're interfacing to some component that requires very fast data rates, in which case you might want to select the fastest option. However, selecting a faster slew rate can increase system noise, so it's not recommended to slew faster unless you really need it.

Differential Signaling

Differential signaling is a method of sending data where you have two signals that aren't referenced to ground, but rather to each other. As I hinted earlier, this is in contrast to *single-ended signaling,* where you have one data signal referenced to ground. Figure 11-3 illustrates the difference.

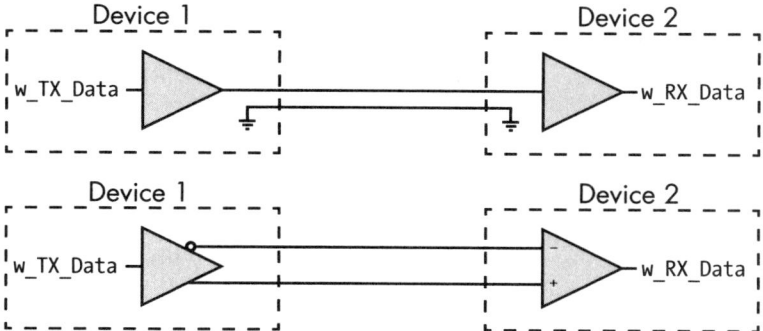

Figure 11-3: Single-ended vs. differential interfaces

The top half of the figure shows a single-ended configuration: we have device 1 transmitting data to device 2 on a single wire, and another wire for the ground path. There's no data on this ground wire, but it's needed to maintain a consistent ground reference between the devices. Data is sent as a voltage on the data wire: 0 V for a 0 or some other value (such as 3.3 V) for a 1, depending on the operating voltage. If we wanted to add another data path, we could just add another single wire between the two devices; the ground reference can work for multiple data paths.

The bottom half of the image shows a differential configuration. Here, we don't have a ground reference passed between the devices. Instead, we have a pair of data lines. Notice the bubble at the start of the upper line, on the output of device 1's transmit buffer. This looks like the bubble we saw when looking at NOT gates and NAND gates back in Chapter 3, and it's a common indication that we have a differential pair. If the difference between the + and – terminals on the receiver is a positive voltage above some threshold, then the signal is decoded as a 1; if the difference is a negative voltage below some threshold, then the signal is decoded as a 0. The details depend on the differential standard. For example, TIA/EIA-644, more commonly called *LVDS (low-voltage differential signaling)*, specifies that there should be a difference of about +/– 350 millivolts (mV) between the two wires. This voltage is quite a bit lower than most single-ended signals use, meaning the system can operate at less power, which is one advantage of differential communication over single-ended communication. A typical FPGA supports about the same number of differential standards as single-ended standards (a dozen or so).

One disadvantage you might have picked up on is that differential communication requires twice as many wires for every data path. In the case of single-ended data transfer, there's just one wire for each data path we want to create. If we want 10 data paths, we need 10 wires (and usually at least 1 ground wire). To create the same 10 data paths with differential signaling, we'd need 20 wires (but no ground wires). This extra wiring costs money and will require a larger connector. Still, differential signals have some unique properties that may make this trade-off worthwhile in certain applications.

One important advantage is that differential signals are much more immune to *noise*, or *electromagnetic interference (EMI)*, than single-ended signals. EMI is a phenomenon caused by changing electrical and magnetic fields—for example, from a nearby microwave oven, cell phone, or power line—that can cause disturbances in other systems. You can think of a wire that carries data as a small antenna that receives all sorts of unwanted electrical signals, creating noise that shows up as a voltage blip on the wire. A large enough voltage blip on a single-ended signal could corrupt the data, causing a 0 to turn into a 1, or a 1 into a 0. With a differential signal, however, the voltage blip will affect both wires equally, meaning the voltage difference between the wires will remain constant. Since it's the voltage difference, and not the exact value of the voltage itself, that matters, the noise is effectively canceled out.

An additional benefit of differential communication is that the transmitter and the receiver can be referenced to different ground voltages and still send and receive data reliably. It might seem strange, but ground isn't always exactly 0 V. The ground in a system can be affected by noise, just as data lines can be, so problems can arise when you rely on ground as a source of truth throughout your system. In particular, it's difficult to maintain a common ground reference for two devices that are far apart, which is why differential signals are often used to send data over long distances. For example, RS-485, a differential electrical standard, can send data at 10 megabits per second (Mb/s) over a distance of nearly 1 mile, which would be impossible with a single-ended signal. Even at closer distances, there are situations where one system might not be referenced to ground at all. Instead, it might be *floating* or *isolated* from a ground reference. To communicate with an isolated system, you need a method of communication that doesn't rely on a shared ground reference; differential communication is one such method.

Differential signals are also able to send data at faster rates than single-ended signals. When a transmitter needs to change from a 0 to a 1, it must drive the line all the way from the voltage corresponding to a 0 to the voltage corresponding to a 1, and that process takes some amount of time (the slew rate). The bigger the difference between the voltages, the more current must be driven onto the line, and the longer the process will take. Since single-ended protocols typically require wider voltage swings between a 0 and a 1, they're inherently slower than differential protocols. For example, the LVCMOS33 voltage swing of 3.3 V is much greater than the LVDS voltage swing of +/- 350 mV. For this reason, almost all high-speed applications use differential signals. We'll get into more detail about this later in the chapter when we discuss SerDes, but interfaces like USB, SATA, and Ethernet all use differential signals for the highest possible data rates.

How to Modify Pin Settings

If you want to specify the operating voltage, drive strength, or slew rate values for your pins, or control which pins are for single-ended signals and which are for differential signals, the place to do it is your physical constraints file. Recall that this file lists how the pins on your FPGA connect to the signals in your design. In addition to specifying the pin mapping, you can also add these other parameters to further define the I/O behavior. Here's an excerpt from a Lattice constraint file that includes some additional parameters:

```
LOCATE COMP "o_Data" SITE "A13";
IOBUF PORT "o_Data" IO_TYPE=LVCMOS33 DRIVE=8 SLEWRATE=FAST;
```

The first line maps the signal o_Data to the pin A13. The second line sets the operating voltage to LVCMOS33, the drive strength to 8, and the slew rate to FAST. You should refer to the constraints user guide for your particular FPGA to see how to set these parameters; the syntax isn't universal across devices. You can also use the GUI in your IDE to set these parameters without having to learn the exact syntax required.

Faster Data Transmission with Double Data Rate

Sending data quickly is where FPGAs can really shine, and one way to speed up transmission is to use *double data rate (DDR)*. Up until this point, I've stated that the signals in your FPGA should be synchronized to the rising edges of the clock. With double data rate, however, signals change on the rising *and* falling edges of the clock. This enables you to send twice the amount of data with the same clock frequency, as shown in Figure 11-4.

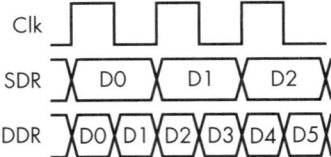

Figure 11-4: Single vs. double
data rate

As you can see, with single data rate, where data is sent on each rising clock edge, you're able to move three bits of data (D0 through D2) during three clock cycles. In comparison, with double data rate, where data is sent on both rising and falling edges, you can send six bits of data (D0 through D5) during the same three clock cycles. This technique is known for its use in LPDDR memory, short for *low-power double data rate*, a type of RAM commonly found in computers, smartphones, and other electronics. Changing the data on both edges of the clock increases the bandwidth of the memory.

You need to create double data rate output (ODDR) buffers anywhere you want to use DDR for data transfer. The details vary between FPGA manufacturers, but I generally recommend creating these ODDR buffers directly within your Verilog or VHDL, using instantiation, since they aren't terribly complicated to configure. As an example, let's take a look at an instantiation template for an ODDR buffer from AMD's Virtex-7 Library User Guide:

Verilog
```
ODDR #(
.DDR_CLK_EDGE("OPPOSITE_EDGE"),
.INIT(1'b0), // Initial value of Q: 1'b0 or 1'b1
.SRTYPE("SYNC") // Set/reset type: "SYNC" or "ASYNC"
) ODDR_inst (
❶ .Q(Q), // 1-bit DDR output
.C(C), // 1-bit clock input
.CE(CE), // 1-bit clock enable input
.D1(D1), // 1-bit data input (positive edge)
.D2(D2), // 1-bit data input (negative edge)
.R(R), // 1-bit reset
.S(S) // 1-bit set
);
```

VHDL
```
ODDR_inst : ODDR
generic map(
DDR_CLK_EDGE => "OPPOSITE_EDGE",
INIT => '0', -- Initial value for Q port ('1' or '0')
SRTYPE => "SYNC") -- Reset type ("ASYNC" or "SYNC")
port map (
❶ Q => Q, -- 1-bit DDR output
C => C, -- 1-bit clock input
CE => CE, -- 1-bit clock enable input
D1 => D1, -- 1-bit data input (positive edge)
```

```
D2 => D2, -- 1-bit data input (negative edge)
R => R, -- 1-bit reset input
S => S -- 1-bit set input
);
```

It's not critical to understand every line here. The most important connection is the output pin itself ❶ ; this is where the ODDR block is connected to the pin. The two data inputs, D1 and D2, will be used in an alternating pattern to drive the data to the output pin. D1 is driven on rising (or positive) edges and D2 on falling (or negative) edges.

Double data rate allows you to speed up data transmission, but if you really want to get your data flying, some FPGAs have a specialized type of interface called SerDes that allows for even speedier input and output. We'll examine this exciting FPGA feature next.

SerDes

A *SerDes*, short for *serializer/deserializer*, is a primitive of some (but not all) FPGAs responsible for inputting or outputting data at very high speeds, into the gigabits per second (Gb/s). At a high level, it works by taking a parallel data stream and converting it into a serial data stream for high-speed transmission. On the receiving end, the serial data is converted back to parallel. This is how the FPGA can exchange data with other devices at very fast data rates. It may sound counterintuitive that sending data serially, one bit at a time, is faster than sending data in parallel, several bits at a time, but that's the magic of SerDes. We'll discuss why this works soon.

SerDes primitives are sometimes called *SerDes transceivers*, which reflects that they can send and receive data. That said, SerDes transceivers are almost always full-duplex, meaning they don't have to switch back and forth between being a transmitter and a receiver like we saw previously with bidirectional communication. You usually set up one data path as a transmitter out of your FPGA, and another as a receiver into your FPGA.

SerDes transceivers help FPGAs excel at sending or receiving very large amounts of data at a rate that wouldn't be possible with other devices. This is a killer feature that makes FPGAs attractive for use cases such as receiving data from a video camera. A high-resolution camera might have a pixel space of 1,920×1,080, with 32 bits of data per pixel, and new images captured at a rate of 60 Hz. If we multiply those numbers, that translates to 3.9Gb of uncompressed raw data per second—quite a lot!—and some cameras can go even higher, up to 4K and 120 Hz. SerDes transceivers allow an FPGA to receive an absolute firehose of data and unpack it in such a way that the FPGA can process it correctly. Another common use case is networking, where you have Ethernet packets flying around at hundreds of gigabits per second. You might have multiple SerDes transceivers working together on a single device to route packets correctly, again at very fast data rates that wouldn't be possible to achieve on a standard I/O pin.

At its heart, SerDes revolves around converting between parallel and serial data. To understand why this conversion is necessary, we need to take a closer look at the differences between serial and parallel data transfer.

Parallel vs. Serial Communication

Parallel communication means we're using multiple communication channels (usually wires) to send data, with the data split up across the different channels. *Serial communication* means we're sending the data on a single channel, one bit at a time. Figure 11-5 illustrates the difference.

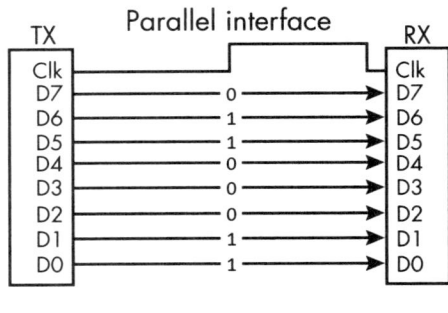

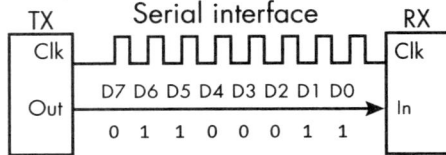

Figure 11-5: Parallel vs. serial interfaces

The top half of the figure shows an 8-bit-wide synchronous parallel interface. *Synchronous* means that we have a clock signal sent between the devices, and the data is aligned with the clock. With this interface, we can send a whole byte of data on a single clock cycle, with one bit on each of the eight wires. In this case, we're sending the value 01100011, or 0x63. While we have eight parallel data paths in this example, you could theoretically create a parallel interface with any width—you could have a 2-bit-wide interface, a 32-bit-wide interface, a 64-bit-wide interface, or any other arbitrary size.

The bottom half of the figure shows the same data transfer of the value 01100011, but it's sent in a synchronous serial stream. Once again, there's a clock signal shared between the devices, but now the data is sent across a single wire, one bit per clock cycle. This way, it takes eight clock cycles to send 0x63, rather than just one clock cycle in the parallel case.

Since parallel communication allows you to send multiple bits in a single clock cycle, it might seem logical that transmitting data in parallel will always allow you to send more data per unit time than sending the data serially. In fact, parallel data transfer runs up against some serious limitations as the bandwidth increases. The physics can't easily scale to support today's high-speed data needs, which is why parallel interfaces are far less common today than they used to be.

If you're old enough to remember the days of ribbon printers, those would connect to your computer using an LPT port, which is a type of parallel interface. Another example was the old PCI bus that was a common way to plug devices like modems and sound cards into your desktop motherboard. Neither of these interfaces is used very much anymore; they couldn't keep up with our need for faster data.

To illustrate why, let's consider how your data transfer speed (or bandwidth) is calculated on a parallel interface like PCI. The first version of PCI operated at a clock rate of 33 MHz and was 32 bits wide, meaning there were 32 individual data wires that needed to be connected between two devices. Multiplying the numbers out, we get 1,056Mb/s, or 132 megabytes per second (MB/s), of bandwidth. This was sufficient for the computing needs of the 1990s, but the demand for more data soon began to increase. We wanted better graphics cards, for example, and the bus to support that data transfer needed to grow accordingly. PCI designers answered the demand by doubling the clock rate to 66 MHz, which doubled the total bandwidth from 132MB/s to 264MB/s. That bought a few years, but it wasn't enough, so PCI next doubled the width of the connector from 32 bits to 64 bits, meaning now we have 64 data wires. This provided around 528MB/s of bandwidth, which again bought a few years, but still it wasn't enough.

By this time, PCI was reaching a point of diminishing returns. There are only two ways to increase data throughput with a parallel interface like PCI: make the bus wider or increase the clock speed. At 64 bits wide, PCI connectors were already a few inches long. To go any wider—say, to 128 bits—the connectors would need to be enormous. Routing all those connections in the circuit board gets very challenging, too. It simply doesn't make sense to continue to widen the bus.

There are also big challenges with increasing the clock speed. When you have a wide data bus and you're sending a clock signal alongside the data, as in the synchronous parallel interface in Figure 11-5, you need to maintain a tight relationship between the clock and the data. The clock will be fed into the clock input of each flip-flop on the receiving side: for a 128-bit-wide bus, for example, there are 128 individual flip-flops that all need that same clock. With so many parallel flip-flops, however, you start to run into problems with *clock skew*, a phenomenon where the same clock arrives at different times to each flip-flop due to propagation delay.

As we discussed in Chapter 7, signals don't travel instantly; rather, there's some delay, and the longer the signals have to travel (that is, the longer the wire) the longer the propagation delay becomes. With a 128-bit-wide bus, the distance the clock signal travels to get to the bit-0 flip-flop can be quite different from the distance the clock signal travels to get to the bit-127 flip-flop. As the clock frequency increases, this difference can create enough clock skew to trigger metastable conditions and corrupt the data.

Clearly, parallel communication has issues. There are fundamental limits to how fast you can go and how many bits you can send at once. Ultimately, the problem comes down to the need to send a separate clock signal alongside the data. The solution is to send the clock and the data together, serially, as part of a single combined signal.

Self-Clocking Signals

Combining the clock and the data into one signal gives you something called a *self-clocking signal*. The process of creating this signal is sometimes referred to as *embedding the clock in the data*, and it's the key technique that makes high-speed serial data transfer via SerDes possible. If you send the clock and the data together as one signal, then the issue of clock skew is no longer a problem, since your clock is received exactly when your data is received; they're the same signal! With clock skew out of the picture, you're able to increase the clock frequency (and therefore the data frequency) tremendously.

There are many different systems, called *encoding schemes*, for embedding the clock in the data. Common ones include Manchester code, High-Level Data Link Control (HDLC), and 8B/10B. We'll focus on Manchester code, since it's a relatively simple encoding scheme, to illustrate one way to create a self-clocking signal. Figure 11-6 shows how Manchester code works.

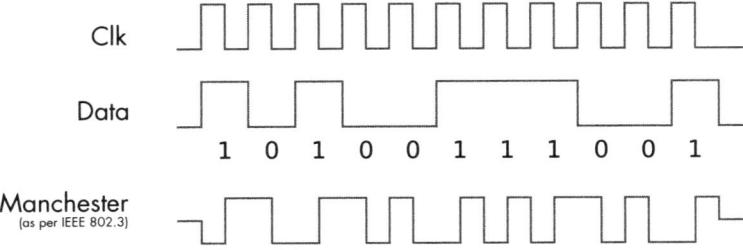

Figure 11-6: Embedding the clock in the data with Manchester code

To implement Manchester code, you take the XOR (exclusive OR) of the clock and data signals, producing a new signal that combines the two. In any given clock period, the data signal will be a 1 or a 0, while the clock signal is a 1 for the first half of the period and a 0 for the second half of the period. The resulting Manchester code signal thus changes halfway through the clock period as well, in response to the transition in the clock signal. Depending on the data value in that period, the Manchester signal will either be low then high (when the data is a 1) or high then low (when the data is a 0). Table 11-2 is a truth table for the Manchester signal based on the different data/clock combinations.

You can use this table to understand the Manchester signal pattern in Figure 11-6.

Table 11-2: Truth Table for Manchester Code

Data	Clock	Manchester encoding
0	0	0
0	1	1
1	0	1
1	1	0

This is simply a truth table for a two-input XOR logic gate, where the inputs are the data and clock signals. As we discussed in Chapter 3, XOR performs the operation either/or, but not both, so the output is high when exactly one input is high, but not when both or neither are high. Looking at the waveforms in Figure 11-6, notice that whenever the clock and data are both high, the encoded value is low. The encoded value is also low when the clock and data are both low, and it's high when only one of the two is high.

The Manchester encoded signal allows you to send the clock and data signals together on a single wire. As mentioned previously, this is the key enabler of high-speed serial data transfer. You no longer need to worry about the alignment of the clock to the data, because *the clock is the data* and *the data is the clock*.

On the receiving side, you need to separate the clock back out from the data, a process called *clock data recovery (CDR)*. This is achieved using an XOR gate and some small, additional logic at the receiver. Then you can use the recovered clock as your clock input to a flip-flop, and feed the recovered data into the data input of that same flip-flop. This way you have perfect synchronization between the clock and the data. The issue of clock skew that we saw with parallel data goes away, enabling you to crank up the data rates far beyond what parallel data transfer could ever achieve.

Manchester code is just one way to generate a self-clocking signal, and it's a simple encoding scheme. It isn't used for modern, more complicated SerDes applications, but it does have some features that are critical. For one, the Manchester encoded signal is guaranteed to transition on each clock cycle. If you're sending a continuous stream of 0s in the data, for example, there will still be transitions in the encoded signal. These transitions are essential for performing CDR on the receiving side. Without guaranteed transitions, the receiver wouldn't be able to lock onto the input stream. It wouldn't know if the data was being sent at 3 gigabits per second (Gb/s), or 1.5Gb/s, or if it was running at all.

Another important feature of Manchester code is that it's *DC balanced*, meaning there are an equal number of highs and lows in the resulting data stream. This helps maintain signal integrity and overcome non-ideal conditions on the wire during high-speed data transfer. We normally consider wires to be perfect conductors, but in reality they aren't; every wire has some capacitance, inductance, and resistance. At slow data rates these don't matter much, but when we get into the Gb/s range, we need to consider these effects. For example, since there's some capacitance in the wire, it makes sense that the wire can be charged up like a capacitor. When a wire becomes slightly charged, for example to a high state, then it requires more energy to discharge it to a low state. Ideally, you don't want to charge up your wires at all: in other words, you want to maintain a DC balance. Sending an equal number of high and low transitions in SerDes is critical to maintaining good signal integrity, and all clock and data encoding schemes have this feature.

How SerDes Works

Now that we've covered the speed advantages of serial communication over parallel communication and examined how to combine the clock and data into one signal, we're ready to look at how SerDes actually works. Figure 11-7 shows a simplified block diagram of a SerDes interface.

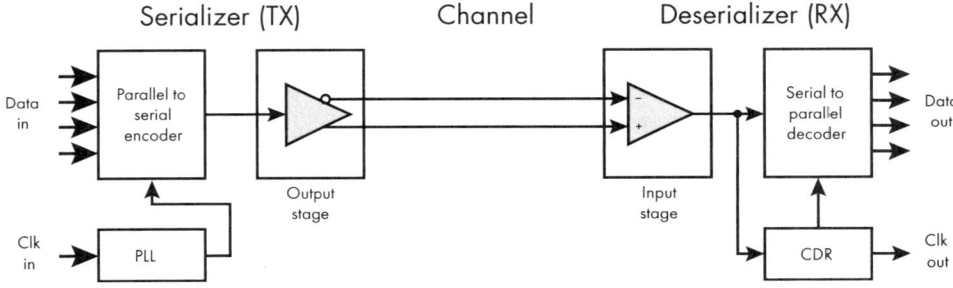

Figure 11-7: A simplified SerDes block diagram

Looking at the image as a whole, we have a serializer on the left that acts as a transmitter, and a deserializer on the right that acts as a receiver. We have clock and data signals going into the serializer on the left, and clock and data signals coming out of the deserializer on the right. That's really all we're trying to do with SerDes: send some data and a clock signal from a transmitter to a receiver. However, doing this at fast data rates requires a lot more functionality than we've seen in simpler input and output buffers.

First, notice that there's a phase-locked loop on the transmit side, in the lower-left corner of Figure 11-7. Usually this is a dedicated PLL specific to the SerDes transceiver that uses a reference clock (Clk In) to generate the clock that will run the serializer. This clock dictates the overall rate at which your SerDes will run. The data that you actually want to send (Data In) comes into the SerDes block in parallel. I've drawn four lines here, but there could be any number of wires. The serializer takes the output of the PLL and the parallel data, encodes it using an encoding protocol, and generates a serial data stream at the desired SerDes rate. For example, if you have a parallel interface that takes 4 bits at a time at a 250 MHz clock rate, then the serializer will generate a serialized version of this data that can be transferred at 1Gb/s, four times that speed.

NOTE *Depending on the encoding scheme, the actual serial stream will likely be running at above 1Gb/s. For example, the 8B/10B scheme takes 8-bit data and encodes it into 10-bit data. It does this for two purposes: to ensure transitions in the clock so we can do clock data recovery at the receiver, and to maintain a DC balance. Going from 8-bit data to 10-bit data adds a 20 percent overhead, however, so to send data at a rate of 1Gb/s we need to send the actual serial stream at 1.2Gb/s.*

The output stage is next. Notice that the output stage contains a differential output buffer. For the reasons discussed previously, such as the ability to send data at high rates, at lower power, and with noise immunity, SerDes transceivers use differential data. The output stage also performs some extra signal conditioning to improve the signal integrity. Once the signal is as conditioned as it can be, it passes through the data channel.

Sending data at high speeds requires optimizations in all parts of the data path, including the data channel itself. For copper wires, the impedance of the material must be controlled to ensure good signal integrity. The channel can also be made from a different material than copper. For example, fiber optics can be used, where light rather than electricity is sent down thin glass or plastic wires. Fiber optics provides excellent signal integrity and is immune to EMI, but it's a more expensive solution than traditional copper wires.

On the receive side, the input stage performs its own signal conditioning to extract the best-quality signal possible. The data is then sent to both a CDR block and the serial-to-parallel conversion and decoder block. The CDR recovers the clock signal from the data stream, and then the deserializer uses that extracted clock to sample the data. It might seem a bit odd that you can recover the clock and then use that clock to sample data from the same signal, but that's the magic of how SerDes works! Finally, at the output side, the data is again converted to parallel. Continuing with the previous example, you would recover your 250 MHz data stream across the four parallel output wires.

This example referred to a 1Gb/s data rate, but that isn't really that fast anymore. As data rates keep increasing, we need to keep optimizing each part of this whole process. Maintaining high signal integrity is critical for SerDes applications. At fast data rates, small resistances, capacitances, and inductances can affect the signal integrity of the line. Passing data through a connector (for example, a USB plug) causes small imperfections in the data path that affect the signal integrity too, so optimizing every aspect of the process becomes critical.

SerDes is one of the killer features of modern FPGAs, but there's a lot of complexity involved. As you've just seen, something as simple-sounding as high-speed data transfer involves a number of steps, including serializing data, transmitting it, receiving it, and then deserializing it again. Even with that process, we're fighting physics to get data to travel at faster and faster rates.

Summary

To be a successful FPGA designer, it helps to have a strong understanding of I/O. This is where the FPGA engineer works at the intersection of electrical engineering and software engineering. This chapter explained how FPGAs use buffers to bring data in and send data out and explored some of the more common settings that FPGA designers need to be aware of when configuring I/O, including the operating voltage, drive strength, and slew rate. You learned about the difference between single-ended and differential communication, and you saw how DDR uses both rising and falling clock edges to send data more quickly. We also explored SerDes, a powerful input/output feature that allows FPGAs to excel at high-speed data applications.

A

FPGA DEVELOPMENT BOARDS

 This appendix lists a few example FPGA development boards that you can use to work on the projects in this book. A dev board that you can program is a valuable learning tool. There's nothing more satisfying than blinking LEDs, pushing buttons, and interfacing to external devices! You can certainly learn a lot from this book without completing the practical examples. But to really unlock its value I recommend purchasing a board, such as one of the devices mentioned here, and working through all of the projects using physical hardware.

In Chapter 2 we discussed some criteria for selecting a board, including the features required to complete the book's projects as written. In particular, I recommend choosing a board with a Lattice iCE40 FPGA, a USB connection, and peripherals such as LEDs, push buttons, and a seven-segment display. The boards covered here either meet these requirements off-the-shelf or allow you to meet them by connecting a few extra peripherals. There are other boards that will work, too; you can use any of these or use these recommendations as a starting point for your own research.

The Nandland Go Board

I created the Nandland Go Board (Figure A-1) as a part of a successful Kickstarter campaign in 2016 to fill a gap in the market: a lack of FPGA development boards that were fun, affordable, and easy to use for beginners. I designed the board with many peripherals, to allow for a wide range of interesting projects. The Go Board has everything you need to work on all of the projects in this book without any modifications.

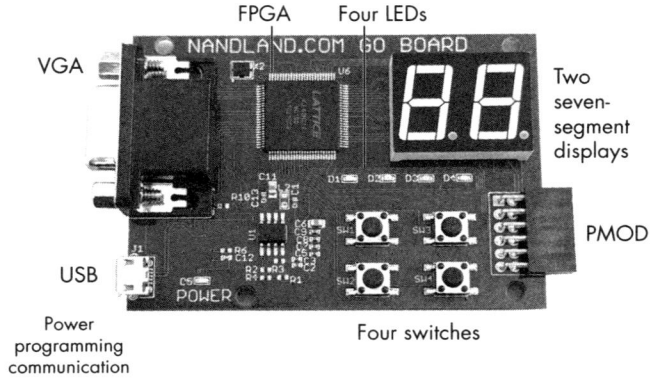

Figure A-1: The Nandland Go Board

The FPGA on the Go Board is an iCE40 HX1K, which is small compared to modern AMD and Intel FPGAs but powerful enough to create the game *Pong* on a VGA monitor. The board has four LEDs, four push-button switches, two seven-segment displays, a Pmod connector, a VGA connector, and a USB connector for power, programming, and communication. At about $65 at the time of writing, it's an affordable board that will allow you to try many different projects in either Verilog or VHDL. It's available through *https://nandland.com*.

The Lattice iCEstick

Lattice designed the iCEstick FPGA development board to be plugged directly into the USB port of a computer, like a thumb drive. It has the same FPGA as the Go Board (an iCE40 HX1K), but it's a bit more limited in terms of built-in peripherals. There are five LEDs available, a Pmod connector, and an IrDA transceiver for sending and receiving infrared data.

At the time of writing the iCEstick is priced at around $50, and it can be purchased directly from Lattice's website (*https://latticesemi.com*) or through electronics distributers such as Digi-Key (*https://digikey.com*). To use this development board for the projects in this book, you'll need to make use of the Pmod connector to expand its capabilities. At a minimum, I would recommend purchasing a breakout board with additional button inputs; for example, Digilent (*https://digilent.com*) sells a Pmod module with four push-button switches.

The iCEstick doesn't have enough Pmod connectors to interface to a seven-segment display and a button module at the same time for the state machine project in Chapter 8. However, if you want to implement the seven-segment display portion of the project, you can connect an individual display using some of the 16 through-hole I/O connections on the sides of the board.

The Alchitry Cu

The Alchitry Cu is similar to the iCEstick in that it's a relatively simple board with a single FPGA and connectors, and not many on-board peripherals. The difference is that the Alchitry Cu has many more connectors available, so you can interface to more peripherals. Additionally, it uses a larger FPGA, the iCE40 HX8K, which has more resources available for larger, more complicated projects.

The Alchitry Cu can be paired with the Alchitry Io Element Board to expand its capabilities: the Io mounts directly on top of the Cu, similar to an Arduino shield, and adds 4 seven-segment displays, 5 push buttons, 24 LEDs, and 24 switches. This option is the most expensive of the three discussed here; the Cu and Io together cost around $85 as of this writing. You can purchase the boards at *https://sparkfun.com*.

Switching Between Boards

As I mentioned in Chapter 2, the beauty of Verilog and VHDL is that they're FPGA-agnostic. The code that you write for one development board will translate very well to another board, often with no modifications needed, provided you aren't using device-specific hard IP features like the ones discussed in Chapter 9. This makes it quite natural to start your FPGA journey on a low-cost, easy-to-use iCE40-based board like the ones described here, and then level up to a fancier development board as you gain experience. You'll be able to take all the projects you developed for your first board and port them to a new, more advanced board with minimal revision.

B

TIPS FOR A CAREER IN FPGA ENGINEERING

Perhaps this book has whetted your appetite for FPGA design to the point where you're considering making a career out of it. Working with FPGAs is a truly rewarding job: you get to wake up each day and solve interesting and relevant problems. Just as designing FPGAs takes practice, applying for jobs is a skill that you can refine. This appendix discusses strategies for getting a great job as an FPGA engineer.

I've been on both sides of the job search process, as both an interviewee and an interviewer, so I've identified some techniques that work to secure a fulfilling job, and I also understand what employers are looking for. In this appendix, I'll share some tips first on how to improve your resume so you can land an interview, then on how to perform well during the interview to boost your chances of getting the job. Finally, we'll discuss how to negotiate for your best possible offer.

The Resume

The purpose of a resume is to get you in the door for an interview. That's really it. Once you're in the door, your resume's work is complete. In my career, I've reviewed hundreds of resumes while trying to fill job positions, and only a small fraction of them have led to an interview. Let's explore some techniques you can use to get your resume to stand out.

Keep It Short

A good engineer knows how to quickly show what's important and what isn't. Demonstrate that by getting straight to the point in your resume. If you have less than five years' experience, there's no reason why your resume should be longer than a single page. Many job seekers think that a longer resume will make them seem more experienced, but padding your resume with fluff will only do you a disservice. Recruiters know filler when they see it. For example, a sentence like "Coordinated with teams from around the world to share responsibilities and tasks" means absolutely nothing. Cut out the fat, and get to the meat.

Instead of including empty platitudes, use your resume to highlight tangible skills, successes, and achievements. If you have no previous work experience, that's fine, but you'll need to demonstrate your qualifications in other ways. For example, describe specific skills you picked up and technologies you learned about in undergraduate classes. If you worked on a large project with a group, that should likely get its own section on your resume.

Include Academic Information

Many companies require a four-year undergraduate degree in engineering as a minimum qualification for a job, so your resume should include information about your education. However, one of the things that I like most about engineering is that it's very much a meritocracy. What counts most isn't which school you went to; it's your knowledge and competency. This means that you shouldn't worry if you didn't go to the most prestigious private school (and if you did, it may not give you the leg up that you expect). When I'm looking at resumes, I usually don't even look at what school the applicant attended. I'm much more interested in seeing evidence of what they've learned and what they've accomplished with that knowledge.

People often wonder if they should include their GPA on their resume. Personally, I would recommend including it if it's above a 3.0, and leaving it off if it's below that. Some companies have GPA requirements (they won't even look at your resume if your GPA isn't listed, or if it's below some threshold), but this appears to have become less common in recent years. Hiring managers understand that being a successful engineer doesn't always correlate with having the best grades in your philosophy class.

If you don't have an undergraduate degree in engineering, your job hunt will be more challenging. Not all companies have a degree requirement, though. Smaller companies, in particular, are often less strict about their minimum requirements. You may be able to compensate for the lack of relevant educational experience by highlighting your practical experience. Be sure to describe projects you've worked on that illustrate your mastery of the field.

Tailor Your Resume to the Job Description

A job description details all the roles and responsibilities that the company expects the new hire to take on. The company is describing their ideal candidate, so you should try to reflect those ideals as much as possible through your resume. This means you should be prepared to tweak your resume for each job you apply for. Highlight areas that match the needs of the company, and consider removing areas that are less of a match. Include relevant buzzwords and specific technologies that you have experience with. Acronyms, in particular, stand out and are eye-catching.

Think about your resume from the hiring manager's perspective. The job description is a list of elements that they're looking for. Don't make it hard for them to find those elements. The more keywords from the job description that show up prominently in your resume, the better. However, keep in mind that you should be able to back up anything you put down on your resume with detailed knowledge during an interview. There's no better way to bomb an interview than demonstrating that you've made false claims of proficiency.

To illustrate how to tailor your resume, let's assume most of your experience has been in Verilog, but you also understand VHDL. An FPGA job description will typically state which language the company uses. If you come across a job that requires VHDL, you should adjust your resume to highlight your VHDL projects. It might even be worth converting some of your Verilog code to VHDL to hone your skills. As another example, the company might want someone with image processing experience. In this case, you should add a bullet to your resume identifying any work you've done in this area. Conversely, if the job posting is from a company building a wireless product, highlight anything you've done with wireless communication instead. Make sure you're specific when highlighting your skills: "FPGA experience with BLE over UART" is much better than "Worked with wireless communications."

Critiquing an Example Resume

Let's look at an example resume for a recent college graduate named Russell Merrick to see what can be improved. Figure B-1 shows my actual resume when I was applying for my first job after college.

Russell Merrick   contact@nandland.com

EDUCATION

University of Massachusetts, Amherst, MA. May 2007

- Bachelor of Science in EE
- GPA 3.630/4.000
- Background in Engineering Management

COURSEWORK

Electronics, Computer Systems, Communications and DSP, Semiconductors, Fields and Waves, Probability and Random Statistics, Management, Marketing, Accounting, Finance.

RELATED WORK EXPERIENCE

Intern, Core IT – Architecture and Planning, Summer 2006

- Designed help documents for wireless home router setup
- Configured routers and IPS Sensors for use in corporate network to prevent against spyware and malicious attacks
- Built servers for use in test lab

Resident Assistant (RA), Fall 2006 - Spring 2007

- Orchestrated the development of a community through programs and events
- Learned conflict resolution by dealing with a range of issues and problems
- Balanced school work with RA responsibilities and engagements

HONORS AND AWARDS

- Vice President of Tau Beta Pi Engineering Honors Society Spring 2005 - 2007
- Recipient of Simon and Satenig Ermonian Scholarship

ABILITIES

Proficient with Microsoft Word, Excel, PowerPoint, Publisher, C++, JAVA, Verilog, Assembly, MySQL, Circuit Maker, Quartus II, AVR Studio, and PSPICE

Figure B-1: An example resume

It wasn't horrible, but in hindsight there's certainly a lot of room for improvement. The biggest problem is that there isn't enough technical content or eye-catching buzzwords. I learned a lot about engineering in college, but based on this resume, you wouldn't know it. I didn't include any details of specific projects that I worked on during my undergrad education that would demonstrate my experience; all I did was list the titles of the courses I took. Additionally, I went into significant detail about non-engineering work that wasn't relevant to the job I was applying for. Let's fix it up.

First, we can remove "Background in Engineering Management" from the Education section. At the time, I thought that having completed some management classes would be an asset, but I quickly realized that as an entry-level employee, you're at the bottom of the totem pole. There's nobody to manage, so your leadership skills aren't relevant. Someday in the future you may advance to the point that you have people working under you, but that certainly wasn't going to be the case in the job I was going for at the time. By the same token, I would remove "Management, Marketing, Accounting, Finance" from the Coursework section. It's just not very relevant for an entry-level FPGA engineer.

In the Related Work Experience section, I describe a brief internship I had working in information technology (IT). This was a valuable experience for me because it taught me one important lesson: I didn't want to work in IT! Here, it would be helpful to use more buzzwords and concrete detail. I worked with Linux computers and Cisco routers, for example, but I didn't write that anywhere on my resume.

The other work experience item, about my time as an RA, could be pared down. It's too long and not especially relevant to FPGA work. The Abilities section can likewise be trimmed. It's taken for granted that job applicants can use Microsoft Word at this point, so don't bother including it. With these cuts, there will be more room to highlight some technical projects I worked on as part of my degree.

Now let's look at an improved version of the resume. The version in Figure B-2 implements these suggested changes.

Russell Merrick   contact@nandland.com

EDUCATION

University of Massachusetts, Amherst, MA. May 2007

- BS in Electrical Engineering
- GPA 3.630/4.000

RELATED WORK EXPERIENCE

Intern, Core IT – Architecture and Planning, Summer 2006

- Managed hundreds of IBM Pentium Xeon Servers via Ubuntu SSH
- Configured Cisco 2800 Routers and IPS Sensors for hacker prevention

UNDERGRADUATE PROJECTS

Analog IR Transceiver, Electronics II

- Built power-amplifier for Infra-red transceiver
- Picked LDO and Switching power supplies for voltage regulation
- Designed Butterworth Low-Pass filter

Microcontroller Motor Control Project, Computer Engineering Lab

- Implemented PID control of Brushless DC motor in C
- UART commands forwarded to Atmel Microcontroller

Resident Assistant (RA), Fall 2006 - Spring 2007

- Learned conflict resolution by interacting with unique individuals

HONORS AND AWARDS

- Vice President of Tau Beta Pi Engineering Honors Society
- Recipient of Simon and Satenig Ermonian Scholarship

ABILITIES

C, C++, JAVA, Verilog, Assembly, MySQL, Circuit Maker, Quartus II, AVR Studio, PSPICE

Figure B-2: An example resume, improved

This revised version is a big improvement. I've added a section detailing a few projects I worked on to illustrate through concrete examples what I learned in college. In describing these projects, I've named many specific technologies that I worked with, like UART, PID, LDO, IR, and so on. Acronyms like these help to grab the attention of the hiring manager. I've also cut parts that were too general, while getting more specific in other sections, such as the description of my IT internship. With this new and improved resume, past Russell could have gotten so many more interviews!

The Interview

Your resume is all about getting your foot in the door, but the interview is where you really need to shine in order to beat the competition and secure the job. An interview is really just an oral test, and as with all tests, it helps to be prepared. You should practice implementing the tips in this section until you're comfortable with them. As awkward as it might feel, find a friend or relative to try out these skills with. It really makes a difference practicing out loud, with another person, as opposed to in your head.

Show Your Enthusiasm

Beyond proving your technical qualifications, an interview is an opportunity to demonstrate your enthusiasm for the job at hand. To do this, it's useful to research the company beforehand. Make sure you understand the company's products, as well as the overall industry they're part of. This way you can impress the interviewers with your level of engagement and your knowledge of the problems the company faces. You might even have some ideas for solutions to a few of these problems!

It also never hurts to contact people at the company directly in advance of an interview. You might think this would be annoying to the people hiring at the company, but it can be hard for them to tell which resumes are submitted by people who are responding to every single job posting they come across, and which people are really excited about the specific open position. I've had job applicants reach out to me on LinkedIn or via email directly, and I'm always impressed when a candidate takes this extra step. It shows they're really interested in the job and driven to achieve their goals.

The most driven candidate I've ever come across literally started a podcast about the space industry to get a job working for my company. He wrote, edited, and published a dozen episodes, just so he could get his foot in the door. I'm not suggesting you need to create a podcast to get a job at your dream company, but just know that nobody has ever been turned away for excess enthusiasm.

Anticipate Questions

As you prepare for an interview, it helps to anticipate the sorts of questions that will be asked. The first thing I recommend doing is reviewing the job description. The topics mentioned there are the ones that are most likely to come up in the interview, as they're the ones the hiring manager is most interested in for that particular position. You probably won't have experience in every single area mentioned in the job description, and that's totally fine. However, I recommend doing some research and learning a bit about any topic you're unfamiliar with, so you at least have some background if that topic comes up in conversation. Make sure you can also speak fluently about anything that you highlight on your resume as a skill.

The job description is always a good place to start, but if you're looking for more ways to prepare, here's a list of some common questions that come up in interviews for FPGA-related positions:

- Describe the difference between a flip-flop and a latch.
- Why might you choose an FPGA over a microcontroller?
- What does a for loop do in synthesizable code?
- What is the purpose of a PLL?
- Describe the difference between inference and instantiation.
- What is metastability, and how would you prevent it?
- What is a FIFO?
- What is a block RAM?
- Describe how a UART works, and where might one be used.
- What is the difference between synchronous and asynchronous logic?
- What is a shift register?
- Describe some differences between Verilog and VHDL.
- What should you be concerned about when crossing clock domains?
- Describe setup and hold time. What happens if they are violated?
- What is the purpose of a synthesis tool?
- What happens during the place and route process?
- What are SerDes transceivers and where are they used?
- What is the purpose of a DSP block?

If I were interviewing someone for an intro-level FPGA job, I would absolutely ask these types of questions. The good news is that each of them has been answered in this book. So study up and crush that job interview!

Pivot

When you don't know the answer to a question, I suggest you pivot by giving the interviewer some information that you *do* know that's relevant in some way. For example, let's say that you only have experience using SVN for version control, and you've never used Git. If the interviewer asks, "How would you create a branch in Git?" don't just say, "I don't know, I've never used Git before." That answer won't be sufficient. Instead, you could say something like, "Well, I haven't used Git for version control yet, but in SVN I use branches to track independent lines of development—for example, when I fix a bug or add a feature. In SVN, this can be done with the command SVN COPY."

This is a fantastic answer. Despite not having the information the interviewer was looking for, you've demonstrated that you understand the purpose of branching and have experience with it. Interviewers get that you won't know all the answers, and that new tools and skills can be learned on the job. Always take the opportunity to explain what you know about a question posed to you, even if you don't know the complete answer.

The Job Offer and Negotiation

At this point, you've beaten the competition and the hiring manager has decided to make you a job offer. Congratulations, you're a professional engineer now! It's time to celebrate. Then, once you've calmed down, I always recommend that you ask for a better offer. You're probably thinking, "But Russell, that's ungrateful. I should be happy that I got any offer at all. What if they pull my offer because I asked them to do better?"

Here's a life lesson that you can take from this book: *it never hurts to ask*. Checking into a hotel? Ask if they have any room upgrades available. Buying something expensive from a store? Ask if they have any coupons or discounts that could be applied. Got your first job offer? Ask if they can increase the salary, stock options, or sign-on bonus. The company will never take back a job offer if you're polite and professional, but every single time I've asked a company for a better offer, they've given it. HR will never give you their best offer first. They expect that a little negotiation will be a part of the process. They want to get you to work for as little money as possible (without insulting you, of course). They always leave money on the table, but it's up to you to get that money. So be polite, but don't be afraid to ask if there's any way that they can increase the offer.

Summary

My goal in writing this book is to spread the knowledge that I've accumulated over my career as an engineer. I hope the book will help you strengthen your FPGA skills and become a world-class performer. Shoot me a message when you've made it. You've got this!

GLOSSARY

application-specific integrated circuit (ASIC)   An integrated circuit that's customized for a particular use, rather than a general use.

bidirectional   Describes an FPGA pin that can both send and receive data, often in half-duplex communication.

blocking assignment   An assignment that prevents the execution of the next statement until the current assignment is executed. The blocking assignment operator is = in Verilog and := in VHDL, but in VHDL you can only create blocking assignments on variables, not signals.

block RAM   A common FPGA primitive used for larger pools of memory storage and retrieval.

Boolean algebra   Equations where inputs and outputs are represented with true/false, or high/low, or 1/0 only.

buffer   An electronic circuit element that isolates its input from its output.

clock   A digital signal that steadily alternates between high and low at a fixed frequency, coordinating and driving the activity of an FPGA.

clock data recovery (CDR)   The process of extracting the clock and data signals from a combined clock/data signal sent through SerDes. The extracted clock signal can be used to sample the data at the SerDes receiver.

clock enable (to a flip-flop)   Labeled En, an input that allows the flip-flop output to be updated when active. When the clock enable is inactive, the flip-flop will retain its output state.

clock input (to a flip-flop)   Labeled >, the input that takes in a clock signal, allowing a flip-flop to work.

combinational logic   Logic for which the outputs are determined from the present inputs, with no memory of past states (also called *combinatorial logic*). Combinational logic generates LUTs within an FPGA.

constraints   Rules about your FPGA that you provide to the synthesis and place and route tools, such as the pin locations for signals and the clock frequencies used in your design.

core voltage   The voltage at which an FPGA performs all of its internal digital processing.

data input (to a flip-flop)   Labeled D, the input to a flip-flop that will be propagated to the output, usually on the rising edge of the clock.

data output (from a flip-flop)   Labeled Q, the output of the flip-flop, usually updated on the rising edge of the clock.

datasheet   A collection of information about an electronic component. FPGAs often have several datasheets, and more complicated FPGAs can have a few dozen.

DC balance   Sending an equal number of high and low bits, in order to improve digital signal integrity in high-speed communications.

debounce   A technique for removing bounces or glitches to get a stable signal. Often used on mechanical switches, which can introduce glitches when a switch is toggled.

demultiplexer (demux)   A design element that can select a single input to one of several outputs.

device under test (DUT)   The block of code being tested by a testbench. Also called the *unit under test (UUT)*.

differential signaling   A method of transmitting electrical signals with two wires by evaluating the difference between them. A reference to ground isn't required.

digital filter   A system that performs mathematical operations on a digital signal to reduce or enhance certain features of that signal.

double data rate (DDR)   Sending data on both the rising and falling edges of each clock cycle. Allows for twice the data throughput compared to single data rate (SDR).

drive strength   A setting that controls the level of source or sink current (in mA) for a pin.

DSP block A primitive used to accelerate math operations for digital signal processing (DSP), in particular multiplication and addition, in an FPGA. Also called a *DSP tile*.

duty cycle The percentage of time a signal is high versus low. Clock signals usually have a 50 percent duty cycle.

edge The point at which a signal, such as a clock, transitions from one state to another. A transition from low to high is called a rising or positive edge, and a transition from high to low is called a falling or negative edge.

edge detection The process of finding either a rising edge or a falling edge and triggering some action based on that edge.

electromagnetic interference (EMI) A phenomenon caused by changing electrical and magnetic fields (for example, from a nearby microwave oven, cell phone, or power line) that can cause disturbances in other systems.

event An action that a state machine responds to, such as a timer expiring, a button being pressed, or some input trigger.

first in, first out (FIFO) A common type of buffer, where the first word written is the first word read out.

flip-flop The critical component inside an FPGA responsible for storing state. Uses a clock as an input and passes the signal from its data input to its data output, usually on the rising edge of the clock. Also called a *register*.

FPGA Short for *field programmable gate array*, a digital circuit that can be programmed with Verilog or VHDL to solve a wide array of digital logic problems.

FPGA development board A printed circuit board (PCB) with an FPGA on it that allows you to program the FPGA and test your code.

frequency The number of cycles (high/low alternations) per second of a signal (such as a clock signal), measured in hertz (Hz).

full-duplex A communication mode between two systems where data can be sent and received at the same time.

generics Used in VHDL to make code more flexible and reusable by overriding behaviors of low-level code at a higher level. Equivalent to parameters in Verilog.

GPIO A general purpose input/output pin, which serves to interface an FPGA to other components on a circuit board.

guard condition A Boolean expression that determines the flow of operations in a state machine. Can be drawn in a state machine diagram using a diamond.

GUI creation The process of using the GUI in an FPGA development tool to create your primitives. This is often the best approach to creating primitives for beginners, as it's the least error prone.

half-duplex A communication mode between two systems where only one system can send data at a time. Also referred to as *bidirectional data transfer*.

hard IP A component within your FPGA dedicated to a particular task, such as a block RAM, PLL, or DSP block. Also called a *primitive*.

high impedance The state of a buffer in which the output accepts very little input current, which effectively shuts the output off and disconnects it from the circuit.

hold time The amount of time the input to a flip-flop should be stable after a clock edge to avoid a metastable condition.

inference The process of creating a primitive using Verilog or VHDL and trusting the synthesis tools to understand your intent.

integrated circuit (IC) Often referred to as a *chip*, a type of electronic circuit in a single package.

linear feedback shift register (LFSR) A special type of shift register that produces pseudorandom patterns by passing certain flip-flops through a logic gate and sending the result back into the input.

logic analyzer A tool used for debugging that analyzes many digital signals at once.

logic gates Devices that perform common Boolean operations like AND, OR, and XOR. Each type of logic gate has a distinctive symbol.

look-up table (LUT) A dedicated component inside an FPGA that performs all Boolean logic operations.

metastability A condition in which the output of a flip-flop is unstable and unpredictable for a period of time.

microcontroller A small computer on an integrated circuit, with a CPU and external peripherals that can be programmed using a language like C.

multiplexer (mux) A design element that can select several inputs to a single output.

non-blocking assignment An assignment in Verilog or VHDL using <=, where these statements execute at the same instant in time. Commonly used to create sequential logic (flip-flops) on an FPGA.

operating voltage The voltage at which a 1 or a 0 appears on a GPIO pin. Common values for a 1 are 3.3 V, 2.5 V, and 1.8 V.

parallel communication A method of transmitting data where multiple bits are sent simultaneously.

parameters Used in Verilog to make code more flexible and reusable by overriding behaviors of low-level code at a higher level. Equivalent to generics in VHDL.

period The time between rising edges of a clock cycle, often measured in nanoseconds. A clock's period is calculated as $1 \div frequency$.

phase The characteristic of a signal that describes the current position of its waveform, or its relationship in time to another signal.

phase-locked loop (PLL) A primitive commonly used as the main clock generator for an FPGA. It can generate multiple clock signals at different frequencies and manage the relationships between them.

physical constraint file A file that maps the signals in your design to physical pins on the FPGA.

place and route The design tool that takes your synthesized design and maps it to physical locations on your specific FPGA. Also performs timing analysis, which reports if a design can run successfully at the requested clock frequency.

primitive A component within your FPGA dedicated to a particular task, such as a block RAM, PLL, or DSP block. Also called *hard IP*.

primitive instantiation Directly creating a primitive FPGA component by using its template. This method allows you to get exactly the primitive you want, without relying on the tools to make any assumptions for you.

propagation delay The amount of time it takes for a signal to travel from a source to a destination.

protocol A system of rules defining how two or more devices communicate.

random-access memory (RAM) Memory that can be accessed in any order, often from one port (single-port) or two ports (dual-port).

register Another word for a flip-flop.

routing Wiring inside an FPGA that gives it its flexibility, but at a higher dollar cost. Also called *interconnect*.

sampling The process of converting an analog signal into a digital signal by taking discrete measurements of it over time.

self-checking testbench A testbench that automatically reports if a design is behaving as expected, without your having to inspect the resulting waveform.

self-clocking signal A signal that uses an encoding scheme to combine a clock and data signal together, such that separate clock and data paths can be merged into a single interface. This technique is essential for SerDes.

sequential logic Logic for which the outputs are determined both from present inputs and previous outputs (also called *synchronous logic*). Sequential logic generates flip-flops in an FPGA.

SerDes (serializer/deserializer) An FPGA primitive used to send data at high rates between a transmitting and receiving device. Parallel data is converted to serial data and embedded with a clock signal before transmission. On the receiving end, the clock and data signals are extracted, and the serial data is converted back to parallel.

serial communication   A method of transmitting data where bits are sent one at a time.

set/reset   An input that, when active, will reset the flip-flop to a default value.

setup time   The amount of time the input to a flip-flop should be stable before a clock edge to avoid a metastable condition.

shift register   A chain of flip-flops where the output of one flip-flop is connected to the input of the next.

sign bit   The most significant bit in a signed number that indicates whether the number is negative (sign bit is 1) or positive (sign bit is 0).

signed   Refers to a signal that can hold positive or negative data.

sign extension   The operation of increasing the number of bits of a binary number while preserving the number's sign and value.

simulation   The process of using a computer to inject test cases into your FPGA code to see how the code responds.

single data rate (SDR)   Sending data on only one edge of each clock cycle, most often the rising edge.

single-ended signaling   A method of transmitting electrical signals where one wire carries the signal, which is referenced to ground.

slew rate   The rate of change allowed for an output signal, usually specified in qualitative terms, such as fast, medium, or slow.

state   In a state machine, a status where the system is waiting to execute a transition. A state can be changed when an event triggers a transition, or if the state itself creates a transition to another state.

state machine   Sometimes called a *finite state machine (FSM)*, a method of controlling the flow through a sequence of operations inside an FPGA.

synthesis   The design tool that turns your VHDL or Verilog code into low-level components within your FPGA, such as LUTs, flip-flops, and block RAMs. Similar to a compiler for a programming language like C.

system on a chip (SoC)   An integrated circuit that combines many components of an electronic system into a single package. For example, an FPGA that has a dedicated CPU might be considered an SoC.

SystemVerilog   A programming language that is a superset of Verilog, with added features that make it useful for verification.

testbench   Test code that exercises your FPGA design code in a simulation environment so you can analyze the design to see if it's behaving as expected.

timing errors   An output of the place and route process that shows signals that might be subject to metastability issues, which could cause your FPGA design to behave unpredictably.

trade study   The act of selecting a technical solution in engineering by examining multiple possibilities and weighing each by its strengths and weaknesses.

transceiver   A device that can both transmit and receive communications.

transition   The action of moving from one state to another in a state machine.

truth table   A table representation of a Boolean equation, listing all possible input combinations and the corresponding outputs.

two's complement   A mathematical operation that converts between positive and negative binary numbers. To take the two's complement, you invert the bits and add 1.

unit under test (UUT)   The block of code being tested by a testbench. Also called the *device under test (DUT)*.

universal asynchronous receiver transmitter (UART)   An interface where data is transmitted or received asynchronously, meaning without the use of a clock. Common for exchanging low data rate information, for example between an FPGA and a computer.

unsigned   Refers to a signal that can hold only positive data, not negative data.

utilization report   An output of the synthesis tool that tells you what percentage of your FPGA resources you've used up.

verification   The process of thoroughly testing an FPGA or ASIC design to ensure it's working as intended.

waveform   A feature of an FPGA simulation tool that shows a visual representation of the signals in your test environment over time.

INDEX

truth tables, 32–39, 41
    AND, 33
    multiple gates, 37
    NAND, 35
    NOT, 34
    OR, 34
    three-input, 37
    XOR, 35
t_{su} (setup time), 133–136
TTL (transistor–transistor logic), 242
Turing, Alan, 33
TWI (two-wire interface), 240
two-dimensional (2D) array, 115, 175
two's complement, 207–208

U

unit under test (UUT), 70–75, 83
universal asynchronous receiver-transmitter (UART), 97–99
unsigned data type, 210
unsigned() function, 211
USB requirements, 15
utilization errors, 125–127. *See also* synthesis: utilization

V

variable keyword, 213
verification, 8, 88–89
Verilog
    background, 9–11
    enumeration support, 152
    weak typing, 10

VHDL
    2008 version, 109
    attributes, 211, 219
    background, 9–11
    data type conversions, 210–211
    strong typing, 10, 178, 182, 210, 213, 217
    verbosity, 22
Visual Studio Code (VS Code), 20
Vivado, 16, 18
voltage, 46. *See also* GPIO

W

wait keyword, 73, 76, 128
waveforms, 74–75, 83–84
when keyword, 61
width, 112, 114
wraparound, 208
write() function, 128

X

Xilinx, 2–3, 16, 69
XNOR (exclusive not or) gate, 36, 99–100, 107
XOR (exclusive or) gate, 35–36, 39, 42, 99, 250–251
xor keyword, 42

Z

Z (high impedance), 239

RESOURCES

Visit *https://nostarch.com/gettingstartedwithfpgas* for errata and more information.

More no-nonsense books from **NO STARCH PRESS**

OPEN CIRCUITS
The Inner Beauty of Electronic Components
BY WINDELL OSKAY *AND* ERIC SCHLAEPFER
304 PP., $39.99
ISBN 978-1-7185-0234-5

BOOK OF I²C
A Guide for Adventurers
BY RANDALL HYDE
440 PP., $49.99
ISBN 978-1-7185-0246-8

ARDUINO WORKSHOP, 2ND EDITION
A Hands-on Introduction with 65 Projects
BY JOHN BOXALL
432 PP., $34.99
ISBN 978-1-7185-0058-7

THE HARDWARE HACKING HANDBOOK
Breaking Embedded Security with Hardware Attacks
BY JASPER VAN WOUDENBERG *AND* COLIN O'FLYNN
512 PP., $49.99
ISBN 978-1-59327-874-8

PRACTICAL IoT HACKING
The Definitive Guide to Attacking the Internet of Things
BY FOTIOS CHANTZIS, IOANNIS STAIS, PAULINO CALDERON, EVANGELOS DEIRMENTZOGLOU, *AND* BEAU WOODS
464 PP., $49.99
ISBN 978-1-7185-0090-7

THE HARDWARE HACKER
Adventures in Making and Breaking Hardware
BY ANDREW "BUNNIE" HUANG
424 PP., $18.95
ISBN 978-1-59327-978-3

PHONE:
800.420.7240 or
415.863.9900

EMAIL:
sales@nostarch.com

WEB:
www.nostarch.com

Never before has the world relied so heavily on the Internet to stay connected and informed. That makes the Electronic Frontier Foundation's mission—to ensure that technology supports freedom, justice, and innovation for all people—more urgent than ever.

For over 30 years, EFF has fought for tech users through activism, in the courts, and by developing software to overcome obstacles to your privacy, security, and free expression. This dedication empowers all of us through darkness. With your help we can navigate toward a brighter digital future.